THE ROBOTS ARE HERE

Dr. Alvin Silverstein
&
Virginia B. Silverstein

Prentice-Hall, Inc.
Englewood Cliffs, New Jersey

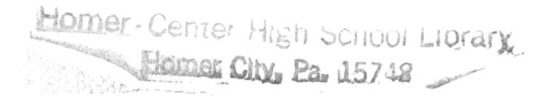

For Glenn and "Mike"

Copyright © 1983 by Dr. Alvin Silverstein
and Virginia B. Silverstein
All rights reserved. No part of this
book may be reproduced in any form or
by any means, except for the inclusion
of brief quotations in a review, without
permission in writing from the publisher.

Printed in the United States of America   ·J

Book design by Constance Ftera

Prentice-Hall International, Inc., London
Prentice-Hall of Australia, Pty. Ltd., Sydney
Prentice-Hall Canada, Inc., Toronto
Prentice-Hall of India Private Ltd., New Delhi
Prentice-Hall of Japan, Inc., Tokyo
Prentice-Hall of Southeast Asia Pte. Ltd., Singapore
Whitehall Books Limited, Wellington, New Zealand
Editora Prentice-Hall Do Brasil LTDA., Rio de Janeiro

Library of Congress Cataloging in Publication Data

Silverstein, Alvin, 1933–
   The robots are here.

   Bibliography: p.
   Includes index.
   Summary: Describes various kinds of robots and their uses and considers the relationship of robots and humans and the implications of a world increasingly run by robots.
     1. Automata—Juvenile literature. [1. Automata.
2. Robots] I. Silverstein, Virginia B., 1937–
II. Title.
TJ211.S54   1983   629.8′92   83-9555
ISBN 0-13-782185-9

ACKNOWLEDGMENTS

The authors would like to thank all those who generously supplied photographs and information for the book. We wish that space had permitted us to use all the excellent photos they provided.

Special thanks to Perry Anable of *Robotics Age*, Charles Balmer, Jr., Gene Beley of Android Amusement Corporation, John Blankenship, Joe Bosworth of RB Robot Corporation, H. R. (Bart) Everett, Jerome Hamlin of ComRo, Inc., Hideyuki Hayashi of Japan Trade Center, Ernst O. Hovland of Hammacher Schlemmer, Kiyoshi Komoriya of Mechanical Engineering Laboratory (Japan), Eugene F. Lally of Robot Shack, Gene Oldfield of Robot Repair, and Glenn Silverstein of Control Automation for their kind help and enlightening insights.

Contents

The Robot Myth	7
What Is a Robot?	17
Robots on the Job	37
Medical Robots	57
Robots in Space	69
A Robot in the House	79
Building Your Own Robot	91
Robots for Fun and Games	102
Robots vs. Humans: Where Do We Go from Here?	111
Glossary	121
For Further Exploration	124
Index	126

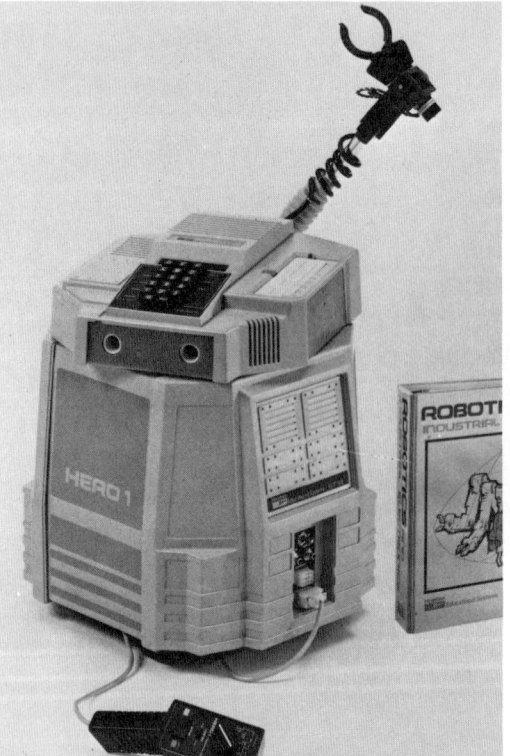

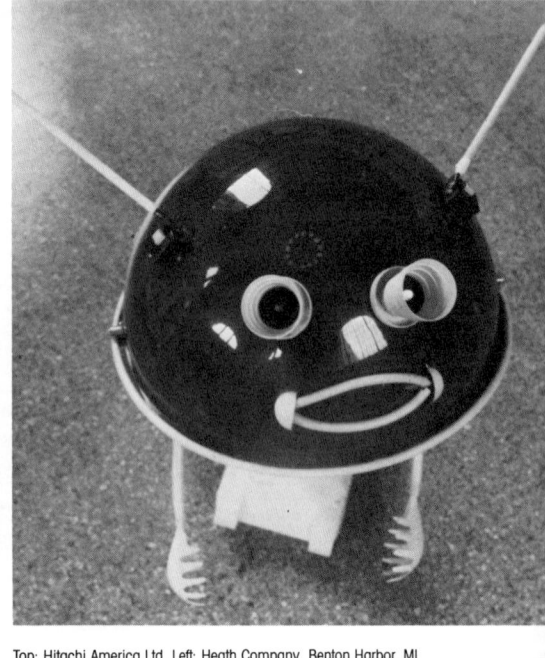

Top: Hitachi America Ltd. Left: Heath Company, Benton Harbor, MI
Right: Courtesy of ComRo, Inc., New York, NY

The Robot Myth

Each December the world waits eagerly to see the first copies of a special magazine cover. Everyone wants to know who will be picked by *Time* magazine as the "Man of the Year." Sometimes the editors choose great heroes who have worked for peace or tried to ease suffering or brought a little more hope or joy to the world. Occasionally they choose a living monster, someone whose evil deeds have had such a deep impact on the world that they cannot be ignored. For weeks the guesses fly. But the *Time* editors' choice in 1982 was a complete surprise. There was no Man of the Year for that year; instead the magazine cover featured a Machine of the Year: the home computer. Computers were sweeping the nation and rapidly changing our lives. Machine of the Year: this was quite an achievement for a tool that only a few years before was still being dismissed as an impractical gimmick for hobbyists.

The ripples of surprise at *Time's* choice had not even begun to die down when another kind of machine made the news. In quick succession, several companies announced that they were ready to sell robots—real working home robots, at a price many people could afford. The headlines shouted, "1983 is the Year of the Robots!" Some articles acclaimed the new devices as the beginning of a new age; others dismissed them, too, as impractical gimmicks for hobbyists. Just a few years earlier, experts were saying that it would be decades before practical, affordable home robots would be possible; some experts are still saying that. Meanwhile, the average person, used to seeing robots performing on TV commercials, on Japanese sci-fi movies, and in person at the local shopping center, might be excused for saying in puzzlement, "What's the big deal about robots? They've been around for ages."

Some confusion is understandable. There are a lot of creatures running around these days that look like robots, act like robots, but *aren't* robots—at least, according to some people's definition. And then there are real working robots that don't look at all like what we normally think of as a robot. As James Albus, a robot researcher at the National Bureau of Standards, remarked a couple of years ago, "There are two illusions about robots. One is that they are already here. The other is that they'll never be here."

Robot myths have been with us for a long time. The ancient Greek gods were fond of making lifelike devices, and sometimes bringing them to life. The popular musical *My Fair Lady* is based on the Greek legend of Pygmalion, a sculptor who created a beautiful statue in ivory and then fell in love with her. Aphrodite, the goddess of love, took pity on him and brought his statue to life. The Greek god Hephaestus made many automata (devices that move about on their own). Hephaestus was the god of fire, and he made women of gold to help him at his forge, as well as tripods that propelled themselves to Olympus and back on golden wheels. Another creation of this mythical god was Talos, a man of bronze who patrolled the beaches of Crete. Talos hurled great rocks at invaders; he could also heat himself red hot and crush trespassers in a fiery hug.

Hephaestus' creations were mythical, but the ancients made some ingenious real automata. Mechanical figures more than 2500 years old have been found in Egyptian ruins. They were statues of the gods that stood in the temples and spoke, gestured, and made prophesies for their worshippers. The ancient Greek architect Daedalus made statues that wheezed, blinked, and scuttled about. One of the ancient writers who had seen the devices commented, "All the works of Daedalus are somewhat odd to look at but there is a wonderful inspiration about them." In China, artisans made a mechanical orchestra back in 202 B.C.

During the Middle Ages, many towns in Europe had huge clocks in the town square. Metal figures came out of the clock to sound each hour. These and other mechanical toys were run by water, steam, or clockwork gears. In the 1600s and 1700s, mechanical automata were a real craze. An Austrian, Joseph Faber, spent twenty-five years building a lifelike mechanical man that could recite the alphabet, ask questions, and laugh. (It spoke English with a German accent.) In the late 1700s, shortly before the Revolution, French royalty were amused by the antics of three automata built for them by Pierre and Henri-Louis Jacquet Droz. One looked like a young boy and wrote letters; another "boy" could draw four different pictures; and a mechanical girl played the piano. They ran by gears inside their chests. The gears could be reset so that the automaton could write a different message, draw a different picture, or play a different song.

A century later, the automaton Psycho amused people with card tricks. Psycho was operated by a bellows, levers, pulleys, and clockwork gears. However, another famous automaton was really a hoax. This was the Turk, a life-sized chess-playing mechanical man built by Baron Wolfgang von Kempelen in 1769. The mechanical Turk toured the world, amazing people with its ability to beat even champion human chess players. Its most famous opponent was the Emperor Napoleon I, who played

Europeans of the seventeenth and eighteenth centuries were enchanted by ingenious clockwork automata like this one.

against it and lost in 1809. Years later, the American writer Edgar Allan Poe matched his wits against the mechanical Turk. The author of two of the earliest detective stories, Poe was more skeptical than most of the people who came to see Baron von Kempelen's automaton. Poe suspected that the desk at which the Turk sat contained a secret compartment large enough to hide a dwarf or a very small man, who could use levers to move the Turk's hand on the chessboard. Poe never got a chance to test his theory personally, but later he was proved correct. Although the artisans of the seventeenth and eighteenth centuries were ingenious, no one at that time could build an automaton that could really play chess, much less well enough to beat a human. (There are robots today that can.)

During the 1800s, automata began to pop up in fiction. E.T.A. Hoffman wrote a story, *The Sandman*, featuring a mechanical woman named Olympia. This story was the inspiration for Delibes' ballet *Coppelia* and Offenbach's opera, *The Tales of Hoffman*. In each of these popular works, a mechanical doll comes to life. A visit to a Swiss museum displaying the automata built by the Droz brothers inspired nineteen-year-old Mary Wollstonecraft Shelley to write a different kind of story. *Frankenstein* was the tale of a dedicated research scientist (the "Frankenstein" of the title) who built an artificial man, but then was horrified at the monster that resulted. Frankenstein's monster was put together artificially, but he was built from biological

The Jacquet Droz figures were among the most famous eighteenth-century automata.

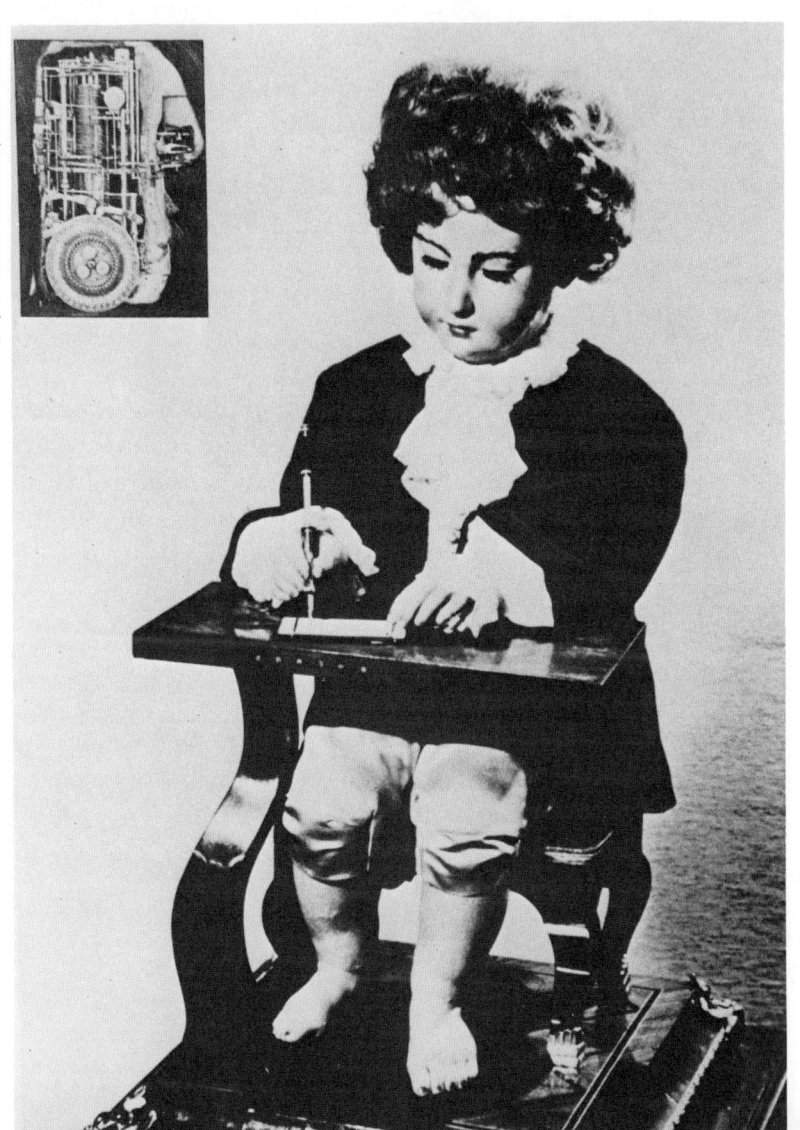

The chess-playing mechanical Turk actually contained a secret compartment to hide a human chess master.

materials. (In science fiction today, he would probably be called an "android," a term that is used for an artificial creature with a basically humanlike form, especially one made from biological materials rather than metal.) Mary Shelley's story dealt with the evil that may sometimes result unpredictably from the search for scientific knowledge, even when the researcher's motives are pure and noble. That is a problem that still troubles many people today, especially in frontier fields like robotics.

Automata of various kinds continued to appear in stories such as *Moxon's Master*. That was a little horror tale written by Ambrose Bierce in 1893, in which an inventor who built a chess machine was unlucky enough to beat his creation in a chess game. The automaton turned out to be a sore loser; it turned on its master and bludgeoned him to death.

Stories like *Moxon's Master* and *Frankenstein* played up the dark side of science and the unfortunate results of trying to imitate life. The automata were often viewed as sinister and evil. That was the basic view of the play that introduced the term "robot" to the world. It was by the Czech writer Karel Capek, and it was first produced in 1921. The title *R.U.R.* was an abbreviation for "Rossum's Universal Robots." Rossum, a greedy factory owner, created a race of artificial men and women to replace his human workers. "Robot" came from a Czech word, *robota*, that can be translated as "worker" or "drudge." Both meanings were appropriate in the play, in which the robots were exploited as slave labor by their ruthless owner. Eventually they rebelled against their human masters and took over the world. But though they were very similar to human beings, Capek's robots had one fatal flaw: they couldn't reproduce. At the end of the play, with humanity nearly wiped out and the robots soon to wear out, a scientist manages to build a male and a female robot, each equipped with that missing spark. They go off together hand in hand, presumably to build a new world and live happily ever after.

The movies made robots even more popular. Most of the movie robots, like those in science fiction stories in pulp magazines and comic books, have been menacing figures of evil, threatening to destroy the world. Despite their superhuman powers, these mechanical monsters were vanquished by brave human heroes.

There have been a few dissenting voices, however. Some writers have seen robots as helpers of humanity. Perhaps the biggest boost to this viewpoint was provided by Isaac Asimov, who wrote a number of robot stories and novels. His first collection, *I, Robot*, was published in 1950 and established the "Three Laws of Robotics":

1. A robot may not injure a human being, or through inaction allow a human being to come to harm.
2. A robot must obey the orders given it by human beings,

except where such orders would conflict with the First Law.
3. A robot must protect its own existence as long as such protection does not conflict with the First or Second Law.

These logical commands were incorporated into the "psionic brain" of each robot in an Asimov story, to ensure that the robots would never turn against their human masters. The difficulties of interpreting the laws (What is "harm"? And how can a robot recognize a true human being?) and the conflicts among the various laws provided interesting plots for the stories. Asimov's robots range from gentle nursemaids and friendly toys, to robot workers that unintentionally become a menace to humans, to a

Robbie, the robot in the movie Forbidden Planet, *was a helpful household robot.*

C–3PO and R2–D2, the famous robots of the Star Wars movie series, are built along very different lines.

sophisticated, humanlike robot detective and, finally, to a sort of computer god that controls the world in the best interests of humankind.

Other science fiction writers quickly adopted Asimov's Three Laws, and benevolent robots began to appear in movies, too. Robbie, the helpful household robot in the 1956 movie *Forbidden Planet*, clearly shows Asimov's influence. Robbie not only does useful things for its human owners, like cooking, cleaning, and manufacturing anything once it has been given a sample, but it is devoted to human welfare. In one scene, Robbie nearly has the equivalent of a nervous breakdown when the scientist Morbius hands it a blaster and orders it to fire at a group of astronauts. (Eventually the scientist saves the robot's sanity by cancelling the order.)

Robot monsters continued to abound in movies and TV: for example, robot villains appeared frequently on the long-running British TV series *Dr. Who* and on the popular American TV and movie series *Star Trek*. But friendlier robots have also been represented. In the 1972 movie *Silent Running*, an idealistic young ecologist tries to preserve Terran plant life in an orbiting dome-shaped habitat. He is aided in his work by three small robots, Huey, Dewey, and Louey, which look very much like the "real" robots that are being built today. Just as friendly but not nearly as realistic are the two famous robots of the movie *Star Wars*, R2-D2 and C-3PO. The squat little R2-D2 served as the model for some of the "personal robots" now coming on the market, but C-3PO is built more on an android plan: he looks just like a metallic man. (In the movie, both characters were played by humans inside robot suits, although a mechanical device was used for some of the R2-D2 scenes.)

Why don't robotics experts consider the clever automata of past centuries real robots? Will we ever have robots like the marauding monsters and helpful housemaids of TV and movies? Is an android a robot, and will we ever be able to build one? What is a robot, anyway?

What Is a Robot?

Joseph Engelberger, founder of the industrial robotics firm Unimation and often called the "Father of Robotics," says, "I can't define a robot, but I know one when I see it."

Anthony Reichelt, founder of Quasar Industries, a firm that makes show robots, says, "A robot is an automated machine with the motor capabilities to duplicate some human motor functions or a shape that emulates the human shape."

Most robotics people might agree with the first part of that definition, though they wouldn't consider it complete, but they very likely wouldn't agree with the second part.

The Robot Institute of America has its own definition, which is accepted by many robot researchers and manufacturers: "A robot is a reprogrammable multifunctional manipulator designed to move material, parts, tools, or specialized devices through variable programmed motions for the performance of a variety of tasks."

Each part of the definition is important. A robot must be a manipulator—it must be able to do things, to move, to produce

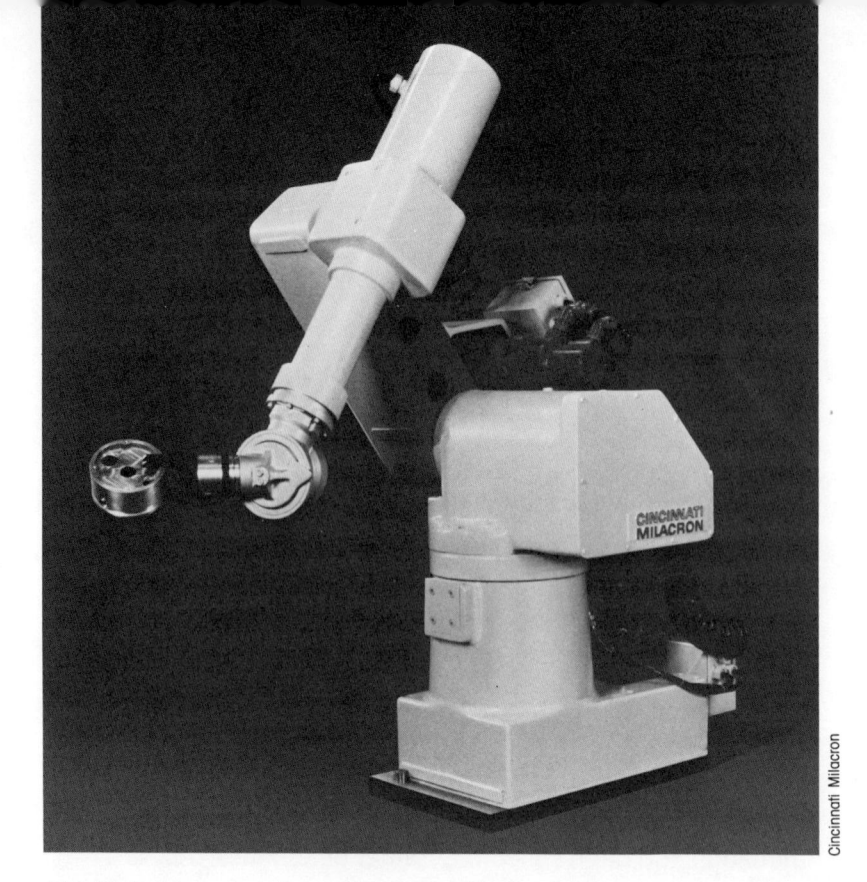

This industrial robot doesn't look at all like the walking, talking robots of the movies.

changes. It is reprogrammable: its instructions can be changed. It is multifunctional and can do a variety of tasks.

That rules out many of the automatic machines at work in factories today: for example, a machine tool that very precisely stamps out holes in metal parts delivered to it by a conveyor belt on an assembly line. It works automatically but it could not be converted to another job—say, spray-painting or spot-welding—if the parts it works on were no longer needed. Notice that there is nothing in the RIA definition that says the robot must be able to move *itself:* a machine can be fixed in place and still be a robot. Nor is there anything that says it must look human: a metal arm with a computer "brain" and a set of replaceable manipulator attachments mounted on a fixed base could still be considered a robot. In fact, that's what many of the industrial robots in use today do look like.

What a far cry from R2-D2 and C-3PO, not to mention the androids of *R.U.R.!* In the real world of robotics today, it's not appearance that counts; it's what the robot *does*.

Although mechanical toys that *seemed* to perform humanlike tasks had been amusing people for centuries, the first real robots were not built until the mid-1900s. They were an offshoot of studies aimed at finding out more about how people and animals think and learn. A mouse or even a worm has a very complex brain. Psychologists thought that they could get new insights into how the brain works by building simple machines that could learn new information and use it to make decisions about their actions. So the first real robots were not mechanical men; they were mechanical animals.

One very simple learning machine, built in the early 1950s, was called Learm. It was modeled after the flatworm, a very popular subject of learning experiments. Learm was worm-shaped, and it received information through three simple switch inputs in its head—one in the center and one on each side.

Learm was placed in a T-shaped maze, like the kind used to train flatworms and earthworms. It moved up the center arm of the T and then bumped its middle switch when it got to the end. This caused it to turn either to the left or to the right. At first the choice was made at random, and Learm was just as likely to choose either side. But then an object was placed in one arm of the T. If Learm turned that way, it bumped into the object with its side switch. The side switches were set up to act as "rewards": bumping one of them triggered an electrical charge in the robot's memory storage on that side. This charge made Learm more likely to pick the same side the next time; the new reward it received increased the electrical charge and made its preference for that side even stronger. Soon Learm was choosing the correct side every time: it had successfully "learned" the T-maze. But if the rewards were stopped, the charge in Learm's memory gradu-

19

ally faded away, and it forgot its lessons. Learm's behavior was actually very similar to the behavior of a real worm in a T-shaped maze.

While some scientists were studying robot worms, others were working with robot turtles. The first two were built by an English psychologist, William Grey Walter, who named them Elmer (*Elec*tro*me*chanical *r*obot) and Elsie (*E*lectro*l*ight-*s*ensitive with *i*nternal and *e*xternal stability). Later models were built at the Aston Cybernetics Laboratory in England and at the Artificial Intelligence Laboratory at MIT. A robot turtle is about the size of a box turtle, with a humped shell about a foot long. It runs on wheels, usually three, and has two motors: one to drive the front wheel and the other for turning and steering. Photoelectric cells and a microphone permit the little robot to respond to light and sound. Researchers have trained their mechanical turtles to move toward a weak source of light and away from a strong one. But the robot can decide to change its behavior, depending on the circumstances: if it is "hungry" (that is, its power is running low), it will move toward the strong light source, where a recharger is kept. Robot turtles have also been trained to respond to sound signals; for example, to stop when the researcher blows a whistle. If the turtle meets an obstacle, it stops, backs off, turns slightly at an angle, and tries again and again until it finally has a clear road. (This may seem like a rather inefficient way of getting around, but it is very much like the behavior of the tiny pond organism *Paramecium.*)

In 1978 two young robot hobbyists, William Hillis and David McClees, formed a company, Terrapin, Inc., to produce and sell Turtle robots. Traveling around a house at six inches a second, flashing lights and beeping, their little robot turtle can learn the floor plan, storing the information in a small computer. It can then draw the floor plan with a pen and vacuum the floor without human supervision. Turtle sold for about $300 and was advertised as the first household robot, but it was bought mainly by computer hobbyists.

The first robot mouse was built in 1952 by Claude Shannon, a pioneer in computer and robot research. His robot mouse could learn to find its way through a maze. Like a real mouse, the robot learned by trial and error, gradually eliminating its incorrect choices until it could move through the maze without running into any blind alleys. Later it could remember the correct path.

Computer researchers soon moved on from the simulation of creatures such as worms, turtles, and mice to efforts to study how humans think. The field of research that tries to duplicate human thought patterns by computer programs is called artificial intelligence, or AI. Today AI is one of the fastest growing and most exciting fields of computer science. AI researchers are

Robot turtles can learn to find their way around. The Terrapin turtles are popular with schools and robot hobbyists. Terrapin, Inc.

trying to work out how we gather and remember information, solve problems, and make decisions. They are trying to build computers that do more than just calculate and follow the instructions of their human programmers. AI programs are designed to make the computers learn from their experiences and generalize them to form new ideas. Such studies have obvious applications to robotics. A robot with a computer brain that can learn and solve problems would be able to do more work on its own, without constant supervision from human controllers.

Before going into the attempts to build "intelligent" robots, let's explore some basics of what a computer is and how it works.

The heart of a computer is its *memory unit*, which stores the information it has received. Computer memory may be stored on magnetic tapes or discs, on small doughnut-shaped rings of a magnetic mineral called ferrite, or on tiny silicon chips. Information is stored in the form of small magnetic charges, produced by flowing electricity. There are two kinds of magnetic charges, which we could think of as "on" and "off," or 1 and 0. It seems incredible that a system of just two symbols, 1 and 0, could be used to perform complicated calculations, store a whole library of words, or provide the instructions for moving a robot arm. And yet, the Morse code—built up from just two symbols, "dot" and "dash"—can be used to spell out any word in the English language.

Computers use what is called a binary code, in which each number is represented by combinations of the two symbols 1 and 0. Our common arithmetic is based on a decimal system, using combinations of ten different symbols. Each number in our decimal system has an equivalent in the binary system. For example:

Decimal	1	2	3	4	5	6	7	8	9	10	11	12
Binary	1	10	11	100	101	110	111	1000	1001	1010	1011	1100

Combinations of the 1 and 0 symbols can also be assigned to letters of the alphabet, just as combinations of dots and dashes are in Morse code. That way computers can produce results in words as well as numbers.

In addition to its memory unit, the computer has some other key parts. There is an *input system,* through which information is introduced. The information may be typed in with a typewriter keyboard, or drawn onto a cathode ray screen with a "light pen." Computers can be built to receive mechanical information directly—for example, the amount and direction of the movements of a robot's arm may be translated into the 1 and 0 computer code and stored in a computer memory. Chemical reactions and light and heat radiations can be used to provide computer input. Speech analyzers have been designed to break down the sound vibrations of speech into patterns of information that the computer can take in.

Another key part of a computer is its *program,* the set of instructions that tell it what to do with the information it has received. The program is written in a special language that is automatically translated into the computer's binary code "machine language." The computer program must spell out every step of the instructions in minute detail. A human being can use guesswork to fill in the gaps in a set of incomplete instructions, but a computer cannot. It does exactly what it is told to do—no more, no less.

The computer's *arithmetic unit* performs the arithmetical operations of addition, subtraction, multiplication, and division. Using the information in its program, a computer can also do other things, such as sorting data and determining whether or not they satisfy a particular set of conditions. For example, a computer could be used to analyze the results of blood tests and determine which ones fall within normal limits and which ones might be an indication of disease.

The early computers, built in the 1940s, were large, complicated devices made of vacuum tubes and a maze of connecting

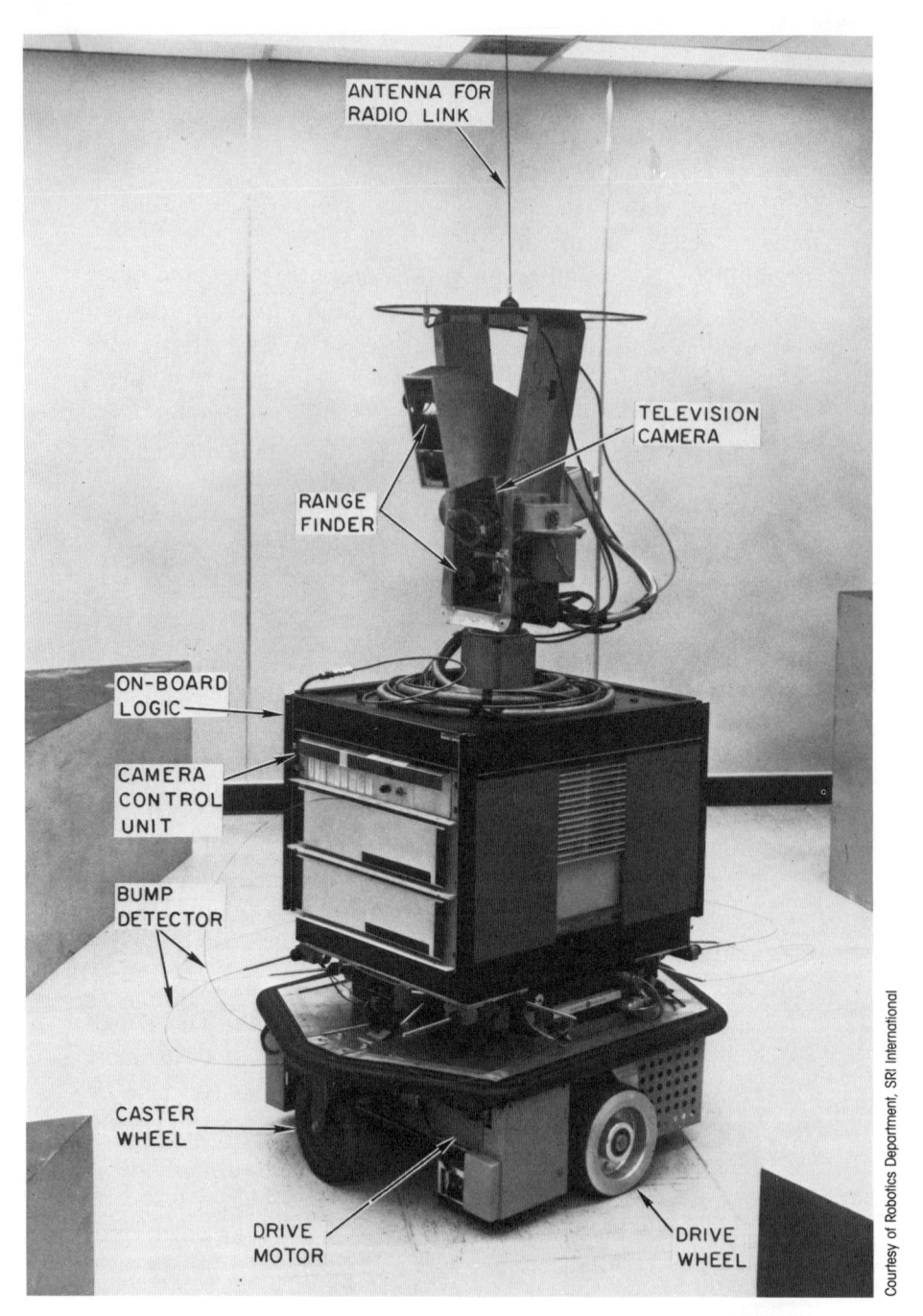

Shakey was an experimental robot designed at Stanford Research Institute.

wires and switches. The invention of the transistor provided a much smaller and more reliable replacement for the vacuum tube. Since then researchers have gone further. Each year computer engineers learn to pack more and more information into tinier and tinier packages. Complicated electrical circuits can be laid out on a tiny chip of silicon small enough to rest comfortably on the tip of your little finger. (The details of the circuits can only be seen under a powerful microscope.) These electronic improvements have made possible an enormous increase in computer power. The early computers were not much more sophisticated than the pocket calculators that can be bought today for five or ten dollars. They could perform perhaps ten or twenty calculations in a second. Today, a home computer can handle as many as 100,000 commands each second, and the world's biggest computers work at a rate of 100 million calculations per second.

In 1968 scientists at Stanford Research Institute in California applied some of the revolutionary advances in computer science to the building of an experimental robot system. The robot consisted of a motor-driven cart that rolled around on wheels, was equipped with bumper switches, and carried a TV camera. Another part of the robot system was not included in the mobile unit: its computer brain was housed in a separate unit that communicated with the roving cart by radio signals. The SRI researchers named their robot Shakey, because it tended to wobble as it rolled around. Peering about with its TV-camera "eye," Shakey could roll down corridors and through doorways, managing to avoid bumping into anything on its way. It followed typed directions like "GO TO POSITION (X,Y)" and "PUSH THE THREE BOXES TOGETHER." By the time Shakey was retired in 1973 (because the government funds to support it had run out), it had learned to solve some rather challenging problems. In one case, Shakey was told to "PUSH THE BOX OFF THE PLATFORM." Shakey didn't have an arm, so the only way to push a box was to roll up next to it and shove. But Shakey moved on wheels, not legs, and it could not climb up the step to the platform. The robot scanned the room with its TV camera and noticed a wedge-shaped ramp lying on the floor. It rolled over to the ramp, pushed it up against the platform, then rolled up the ramp onto the platform and pushed the box off.

Since Shakey's time, the enormous advances in computer science have made it possible to build much smaller and more powerful "electronic brains." Now a robot's computer brain can easily be packed inside its body, although this is not always done. (In some factories, for example, a number of robots may share the same computer brain, which operates them by remote control.)

Today's robots are instructed to do their tasks in an operation called programming. In some cases the operator actually "walks" the robot through its new tasks in slow motion, while the robot's computer brain notes and memorizes each movement in the sequence. (Some robots can be taught to perform a sequence of as many as 2000 separate steps.) Often a robot is programmed with a series of instructions typed out on a keyboard mounted on it, or on a small portable keypad, called a teaching pendant, that communicates with the robot's computer.

The sequence of commands involved in a seemingly simple movement like swinging a robot arm over to grasp a machine part may be amazingly long and complex, since it must specify the movements of all the joints in the arm and their precise positions in space. Robotics researchers are constantly working to incorporate more of these instructions into the robot's built-in programming, so that the instructions from the human programmer will not need to be so complete and precise. For example, Intelledex, an Oregon robotics firm, has developed a programming language called Robot Basic. The Intelledex researchers started with a form of Basic, the programming language used in personal computers, and added 100 command verbs specially suited to the movements of a robot manipulator.

Some researchers are developing ways of controlling robots

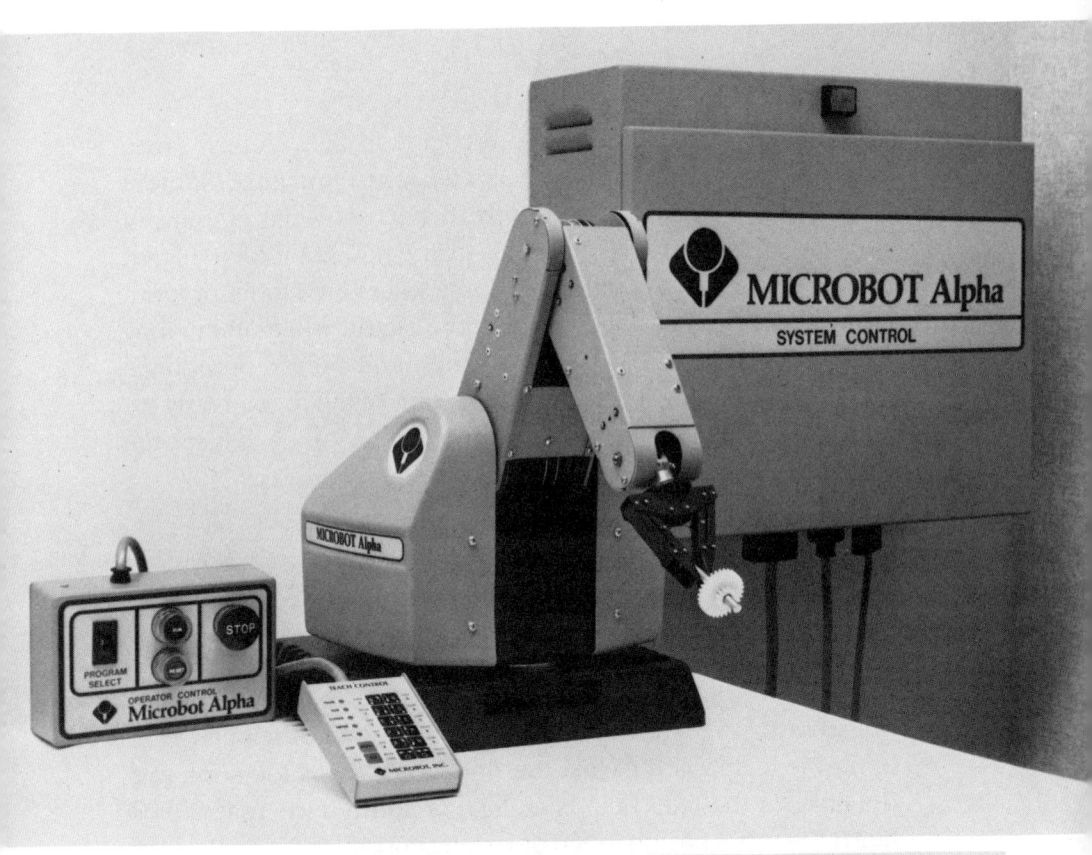

Peter Nuding, Courtesy of Microbot, Inc.

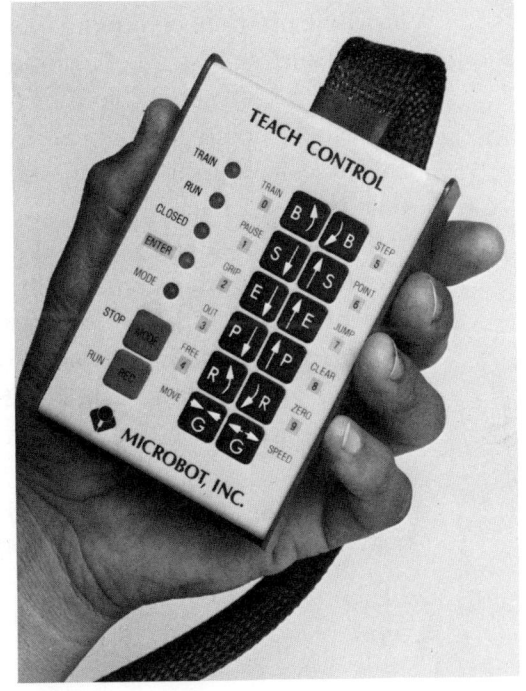

A small teaching pendant may be used to control a robot's movements.

with spoken commands. Voice recognition devices have been invented, but they are still fairly limited. The number of different commands a robot can recognize depends on the amount of computer memory devoted to this task. Another problem is that people's voices sound different. Some voices are higher or lower than others; some people speak more slowly, while others may slur their words or run them together; and people pronounce words differently. Usually a robot can respond correctly to spoken commands from only one person; it needs special retraining to be used by someone else.

Robot senses are another very active area of research and development. Picture a robot whose job it is to screw light bulbs into an automobile dashboard. The robot's memory contains instructions for all the movements involved in picking up a bulb and screwing it into a precisely placed hole in the dashboard. But what happens if the conveyor system that is supposed to bring the dashboard to just the right place stops too soon and the hole for the light bulb is an inch or so from where it should be? If the robot has no touch or vision sensors, it will go through its whole program of motions, vainly trying to screw the bulb into a solid surface instead of moving it over to where the hole really is. If the robot's supply bin runs out of bulbs, it would act equally "foolishly," going through all the appropriate screwing movements while its gripper fingers grasp empty air instead of the bulb that isn't there.

Robotics researchers are working on a variety of sensors to give robots information about their surroundings and help them to act more intelligently. Scientists and engineers use the term *"feedback"* for information about the results of an action that is used to modify future actions. Equipping robots with sensors provides feedback that can make their work much more effective.

Some robot touch sensors work like cats' whiskers, signalling contact with a solid object and helping the robot to move around without bumping into walls and other obstacles. Researchers at MIT are working on a sensitive "artificial skin" to give robots a sense of touch. The skin consists of thin layers of rubber, lined with wires. When the robot touches an object, the layers are pressed together, permitting more current to flow through the wires. The greater the pressure, the more current flows. By reading the patterns of electrical current, the robot's brain can put together a picture of the object it is touching. A robot hand with this sensitive skin can pick up a paper cup without crushing it.

Robot vision systems in use today are quite different from the vision system we humans use. The robot systems use a TV camera that divides the field of vision into a series of horizontal slices that it rapidly scans across, one line after another, left to right, and then moves down to the next, until the whole field has been covered. The robot's computer brain divides each line of the picture further into a series of points. Each point on the picture is called a pixel, short for "picture element."

The computer's program sets a threshold level for brightness: any pixel that is brighter than that level will rate as 1, and any pixel that is darker than the threshold will rate as 0. When the robot's brain puts all the pixels together to form the whole picture, objects contrasting against the background will show up in a pattern of 1's and 0's. But the picture the computer "sees" is very different from the view our human eyes give. It is a flat two-dimensional image, not the 3D image we see. What is more, the robot's picture is only a silhouette—no shades of light and shadow, only the flat contrast. (In more sophisticated vision systems, several thresholds may be used. The pixels may be rated, for example, on a scale from 0 to 5, where 0 is white, 5 is black, and 1 to 4 represent shades of gray. The picture is more realistic, but it is still two-dimensional.)

In some robot vision systems, the computer is programmed to register only areas where the light intensity changes sharply. That way it picks up the edges of objects and draws their

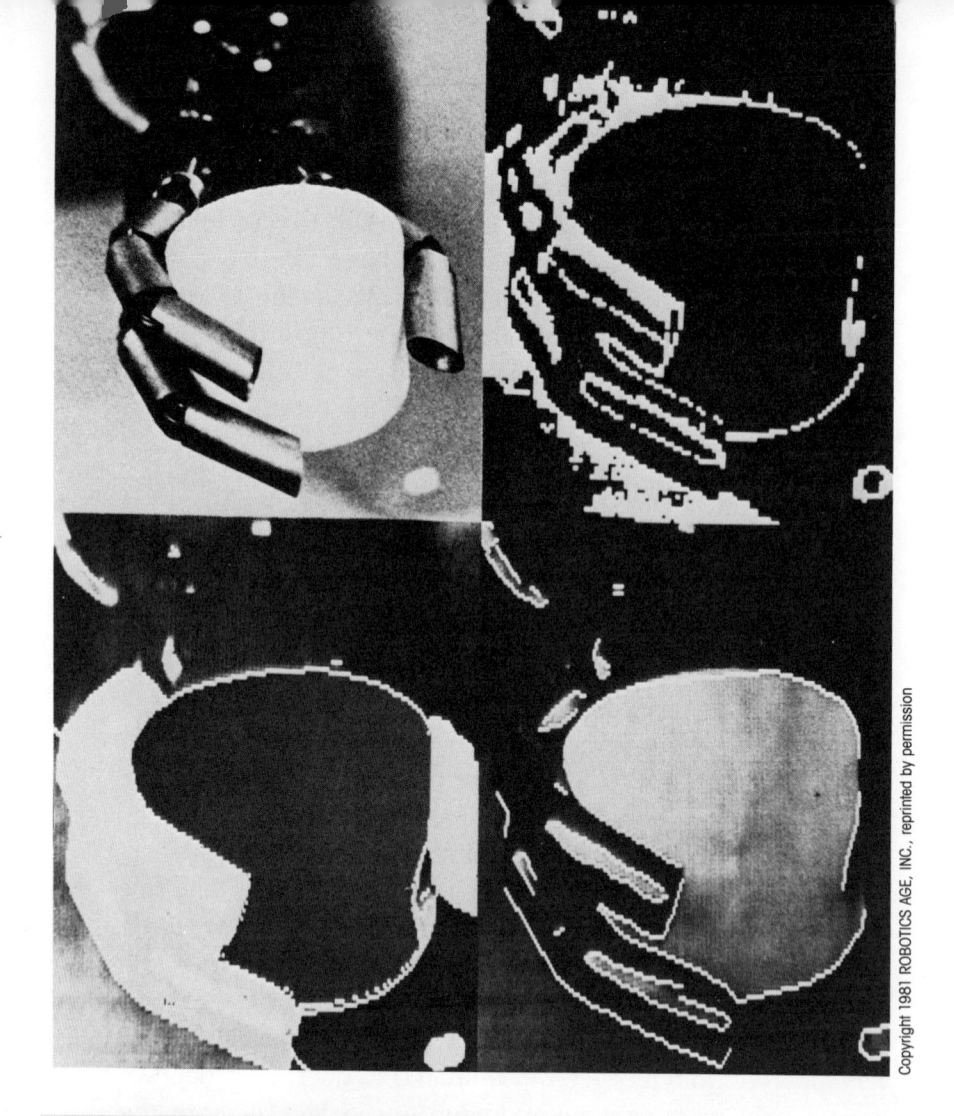

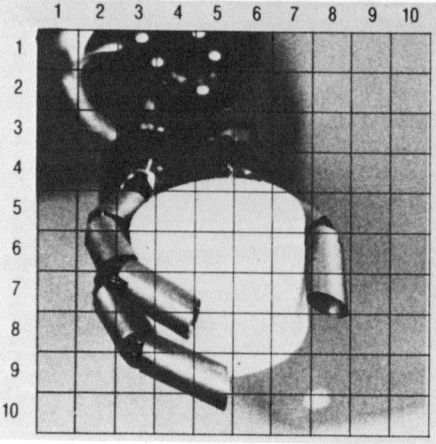

In successive views (above), robot vision systems get a better defined picture, but they are still far from the clarity and detail of the image at upper left. The system divides an image into rows and columns (left) and assigns a number to each box (pixel) according to its brightness. (In an actual system, the divisions would be much smaller.)

outlines. This kind of vision system is useful for many jobs, such as inserting screws into holes and inspecting machined parts to make sure that they have been shaped accurately.

The crude robot vision systems that have been developed so far are good enough for many useful applications. Octek Inc. has developed a robot arm with a built-in video eye that can examine and identify the caps for typewriter keys, pick them up, and set them in place on the keyboard. Intelledex's 605 manipulator can memorize the shapes of 100 different parts. In the learning

The Intelledex 605 robot can memorize the shapes of 100 different parts. Here it is loading an odd-shaped component into place on a printed circuit board.

process, each part is shown to the built-in camera in five different positions; then the 605 can distinguish between tiny parts such as resistors, diodes, and capacitors.

Picking up and sorting parts, however, is a very limited kind of task. Many jobs that humans do with ease require dealing with objects that move around and shift positions and angles. To function effectively in this "real world," robots will need a more sophisticated vision system. Researchers are working on combinations of computers and TV cameras that produce 3D images to give robots depth perception. But so far these systems are too slow. It may take as long as several minutes to analyze a single picture. By the time the robot has finished its computations, everything may have moved. An effective vision system must tell the robot what is happening *now*, not a few minutes ago. As computing power gets faster and cheaper, such "real-time" vision systems may soon become possible.

Control Automation's Mini-Sembler robot works on printed circuits, calculators, computer parts, and small motors. It can position objects to within 1/1000 of an inch in less than 5 seconds.

Left: Sato Akira, Mechanical Engineering Laboratory, MITI, Japan. Right: Photo courtesy of Odetics, Inc.

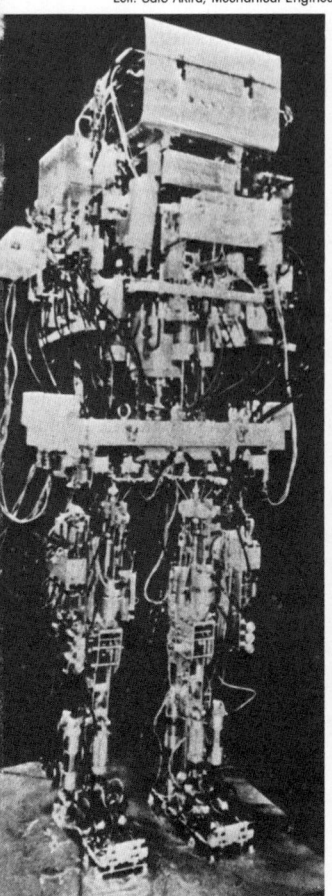

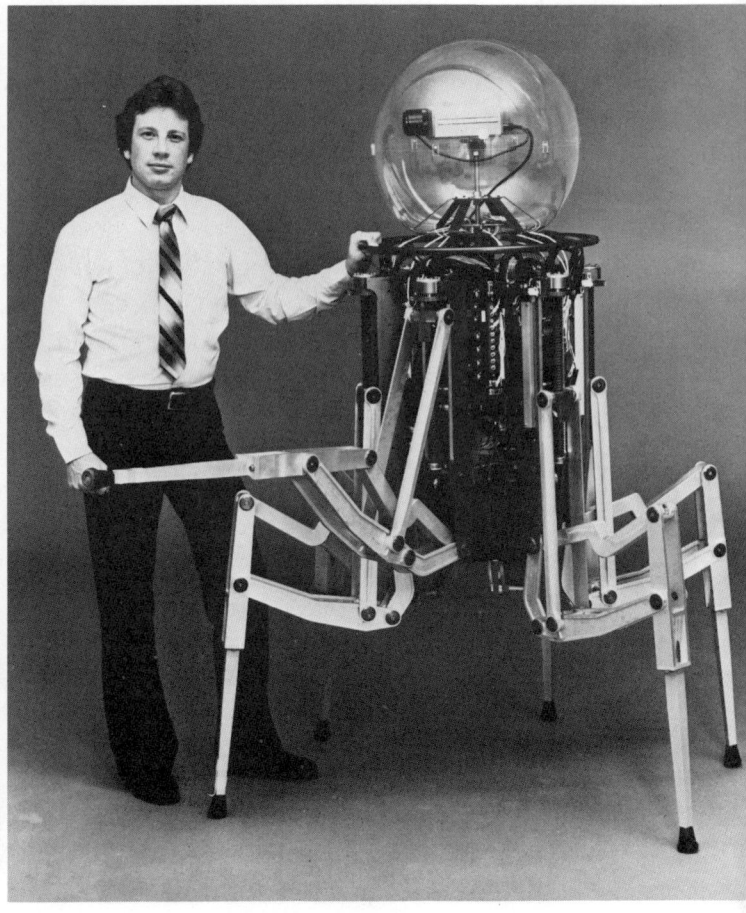

Two variations on a walking robot: an experimental two-legged walking machine designed in Japan and the six-legged ODEX−1.

Roboticists are steadily improving the capabilities of robot manipulators. The simple grasping claw of the early models seems crude in comparison with the three-fingered "Hi-Ti" hand, produced by Hitachi Limited of Japan. This robot hand can wiggle parts together when they are a bit out of line, and it is even agile enough to twirl a baton. Robotics engineers at the Charles Stark Draper Laboratory in Cambridge, Massachusetts, designed a flexible wrist for jobs that involve inserting a peg-shaped object into a hole and installing screws. A robot equipped with the Draper wrist can assemble an automobile alternator in just 65 seconds—a job that takes 90 seconds for an experienced human worker.

For many people, robots won't truly be "real" until they are able to walk around and talk. Most of the mobile robots in use today move around on wheels, but researchers are working on designs for walking machines. The action of walking is far more complicated than it seems. In addition to the coordinated physical movements of the legs, the brain controlling them must take into account a whole delicate balancing act involving the rest of the body. (The next time you are walking down the street, try to notice how much of the time you are actually standing on only one foot and shifting your weight, getting ready to transfer it to the other foot. You have learned to do it automatically without thinking about it, but a robot's program would have to provide for every tiny movement along the way.)

Researchers are making computer studies of how people and animals walk, trying to devise systems to help paralyzed people move around. Their research will also be helpful to roboticists developing walking systems for robots.

In 1983, Odetics, Inc., in Anaheim, California, demonstrated ODEX-1, a remote-controlled six-legged walking robot. The choice of a six-legged design was a shrewd one: studies of how animals move have shown that six-legged insects have much greater stability than four-legged or two-legged animals. Even with three legs off the ground, an insect (or an ODEX-1) still rests on a sturdy tripod. The movements of the robot's legs are controlled by seven microcomputers, one for each leg and one to coordinate the rest. These computer brains receive a flow of information about the positions of each leg relative to the others and to the ground, so that the human operator can make ODEX-1 move using two simple joysticks, rather than having to command each leg separately. The robot can lift and carry up to 1800 pounds and will be used in undersea mining, nuclear plant work, and other jobs too dangerous for people or in places so rough that only a walking machine can move around effectively.

As for talking robots, they're well on the way. Speech synthesizers are not only available, they are getting so cheap that manufacturers are building them into toys. There will soon be many more talking machines. In about 200 supermarkets in various parts of the United States, talking cash registers announce the price of each item, the total, and the change due the customer. Talking Coke machines greet customers, ask them to "Make your selection, please," and remind them, "Don't forget your change."

Reactions to the talking machines have been mixed. Many customers like them and find them friendly, but others think they are a waste of money. Clerks and other workers who must listen to a rather monotonous computer voice saying the same

The voice synthesizer in a Kurzweil machine reads a book for this blind student at the New York Public Library.

things over and over all day tend to be very unenthusiastic about the idea.

But as the voice synthesizers and other microprocessor applications become cheaper and more effective, talking machines will probably be even more common. Speech synthesizers in elevators, prompted by appropriate sensors and a built-in computer program, will warn passengers to "Please stand clear of the doors," or "Please put out your cigarette." Talking dashboards are already being incorporated into cars; soon they'll be nagging us to buckle our seatbelts and warning us when the gas is getting low. Blind people can place a book in a Kurzweil machine and listen while a synthetic voice reads the pages.

FM radio disc jockeys in Pittsburgh recently used a text-to-speech machine nicknamed "Hal" as the announcer for an entire radio show. Few of the listeners realized that it wasn't a human talking, as "Hal" gave the news, sports, and weather, told some jokes, did the lead-ins for the records, and ended the program with the line, "E.T., phone home." Each sentence spoken by the machine had been typed into a computer programmed to convert the words to phonemes, the sound elements of speech. The machine strung the phonemes together to form realistic-sounding words and sentences. Humans gave "Hal" his lines, sentence by sentence, but speech programs for robots can be combined with speech recognition to permit the robots to hold realistic-sounding "conversations" with people.

A talking robot doesn't really understand what either it or the person conversing with it is saying. And so far, even the robots designed to "think" have nothing approaching the mental sophistication of a human being. Will we ever be able to build robots as intelligent as people—robots capable of holding a real conversation or coming up with an original idea? Maybe, and maybe not. First, let's find out more about what today's robots can do.

Robots on the Job

Clyde is one of the hardest workers at the Ford Motor Company stamping plant in Chicago. He's popular, too. His fellow workers all call him by his nickname, Clyde the Claw. When he had a breakdown, they sent him cards and flowers and threw a get-well party for him. Clyde isn't a human being. He is an industrial robot, and his real name is Unimate 4000B.

In industry, robots are no longer science fiction; they are practical reality. Sales of robots in the United States amounted to about $125 million in 1981. Though there were fewer than 7000 industrial robots working in U. S. factories in 1982, their number is expected to grow to more than 100,000 by 1991. In Japan, robots are more widely used: estimates range from 14,000 to more than 77,000 already installed there. (The wide range is due to disagreements over the definition of "robot": the Japanese count many devices that U.S. roboticists would consider just "machines," because they are manually controlled or are not reprogrammable.)

Robots bring enormous savings of time and money and great

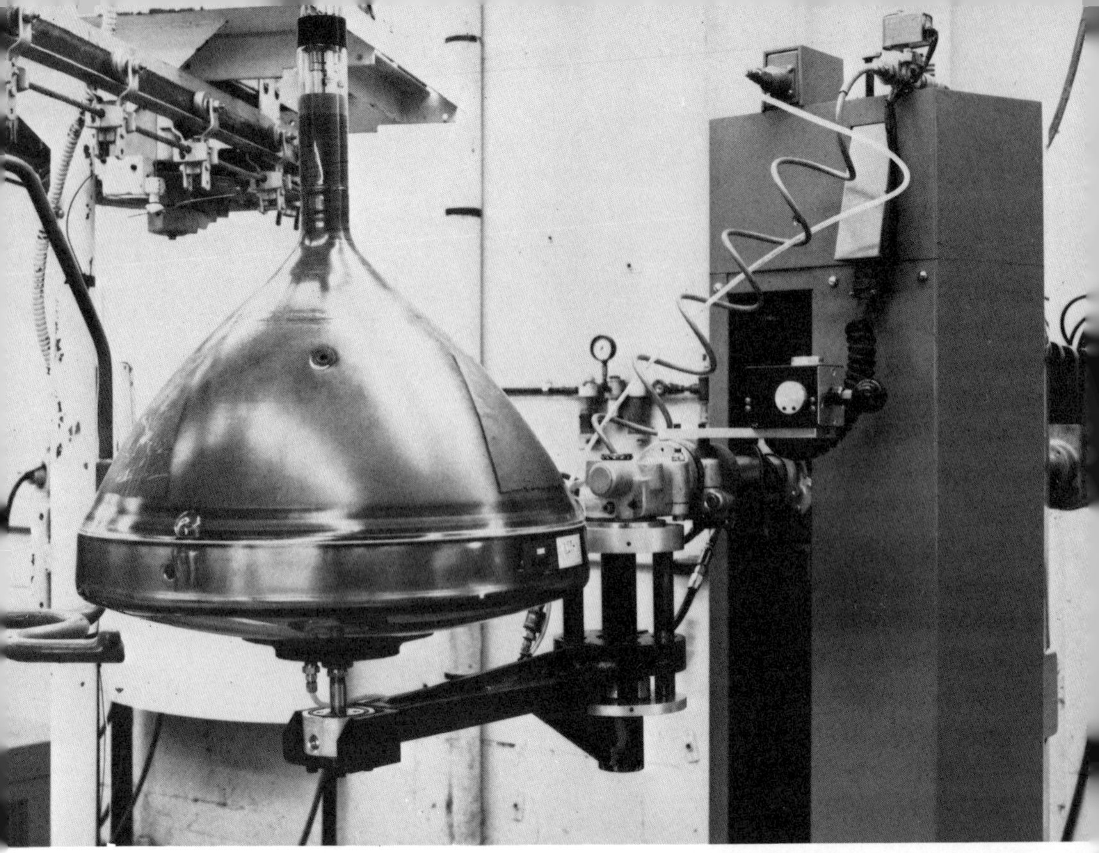

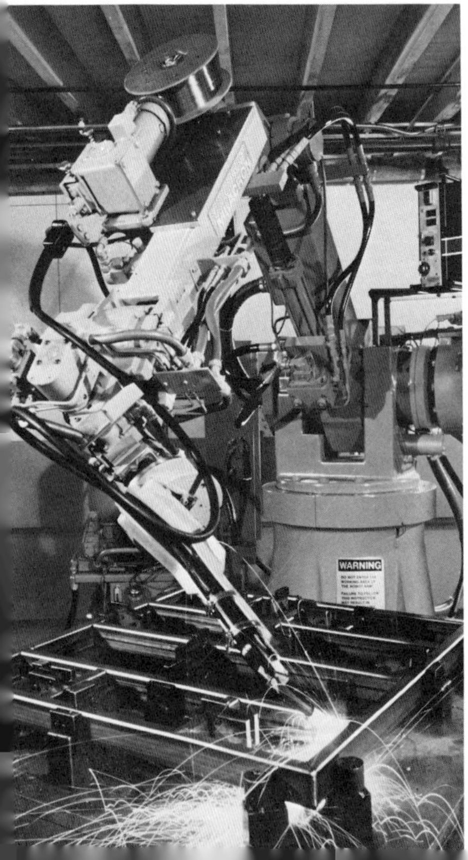
WARNING

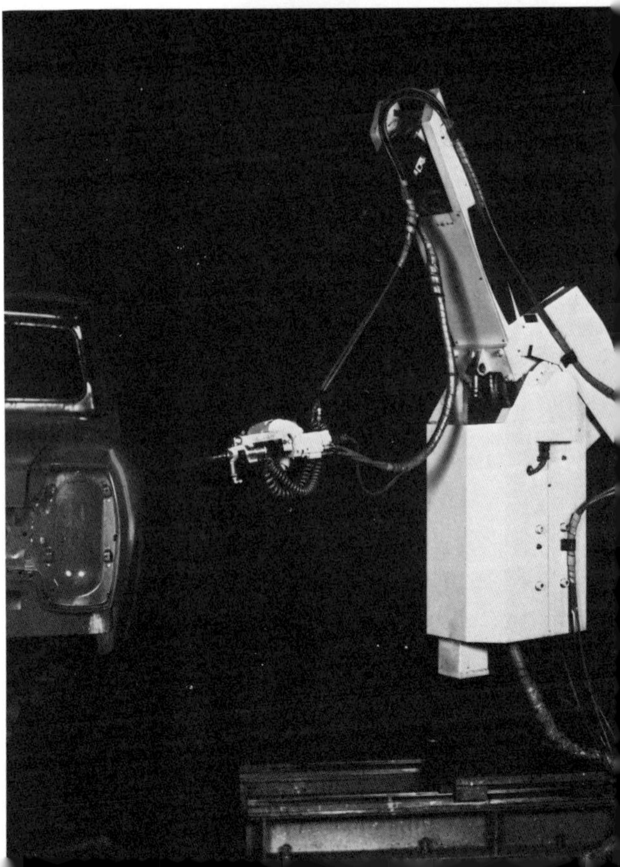

increases in productivity. After the initial cost of buying a robot, a company doesn't have to worry about paying it a salary or fringe benefits, as it would to a human worker. A robot doesn't take coffee breaks, and it doesn't even need to go home and sleep. It can keep right on working, twenty-four hours a day, seven days a week. It can keep up this tireless pace for years with only minor maintenance: the industrial robots in use today typically can work for about 40,000 hours before they need an overhaul.

What is more, robots can do jobs that humans can't do or don't want to do. Robots excel at tasks that are hard, dirty, boring, or dangerous. In his *Manager's Guide to Industrial Robots*, robot manufacturer Ken Susnjara advises a manager thinking about installing robots to take a walking tour through the factory, looking for certain telltale signs. Look for crowds, he says. Wherever there are a lot of workers all crowded into one spot, robots may be able to ease the workload. Look for "machine people"—human workers doing boring jobs that don't require any thinking or judgment. Those jobs are perfect for a robot, and the humans would be happier doing something more challenging. Look for "closets with skeletons"—parts of the factory that a manager would be embarrassed to show to an important visitor, such as those with hot, dirty, or dangerous working conditions. Look for "ridiculous processes," Susnjara advises—operations that would look strange or silly to an outsider coming into the plant for the first time. For example, workers on spray-painting jobs must often work in spacesuits—completely enclosed suits with outside air pumped in. The workers can't breathe the air of their workrooms, because the fumes from the sprays would be unhealthy or downright poisonous. Working in a spacesuit is

Robots on the job: a Prab model E robot that can handle parts weighing up to 100 pounds (top), a Cincinnati Milacron T³ welding the base for a computer main frame four times as fast as a human welder can (left), and a Hitachi spray painter (right).
Top: Prab Robots, Inc. Left: Cincinnati Milacron. Right: Hitachi America Ltd.

awkward and uncomfortable. But a robot wouldn't need a spacesuit, because it doesn't need to breathe.

Other good candidates for robot jobs include operations where the company is thinking of buying an expensive single-purpose machine (a reprogrammable robot might be able to do that job and others, too) and operations where a lot of human labor is wasted on redoing jobs that didn't come out right (a robot can do it right the first time, every time). And finally, Susnjara says, look for jobs you wouldn't want to do yourself—jobs that you would find too difficult, tiring, or boring. A robot wouldn't mind them at all.

The first working industrial robots were developed in the late 1950s for work in nuclear power plants. Some areas of a nuclear plant are radioactive, and it is too dangerous for humans to work there. Remote-controlled devices called Mobots (short for Mobile Robot) solved the problem. Long flexible arms with gripper "hands" could do complicated and delicate jobs, like pouring radioactive liquids from one test tube to another. Television cameras mounted on movable stalks acted as "eyes" for the device, providing a picture that could be watched by its human controller.

Manipulators, controlled either directly or remotely by human operators, are still used for many jobs, from underwater exploration and salvage to the handling of tiny laboratory animals in germfree experiments. But today most people would not consider them true robots.

The first of the modern industrial robots was the result of a conversation in 1956. An inventor named George Devol had an idea for a "universal manipulator" and was trying to find some backers with enough money to help him develop it. At a party, someone introduced him to aircraft engineer Joseph Engelberger. Engelberger had long been an avid science fiction reader, and when he heard Devol's ideas, it was like Asimov's *I, Robot* stories coming to life. That chance meeting was the start of a long and fruitful working relationship. Devol had the creative

Courtesy Masasumi Miyakawa,
Nagoya University School of Medicine, Japan

Remote-controlled manipulators still have their uses. Here a researcher works the manipulators on a germfree rearing unit. Inside the chamber, the mechanical "arms" handle animals that have never been exposed to the germ-filled outside world.

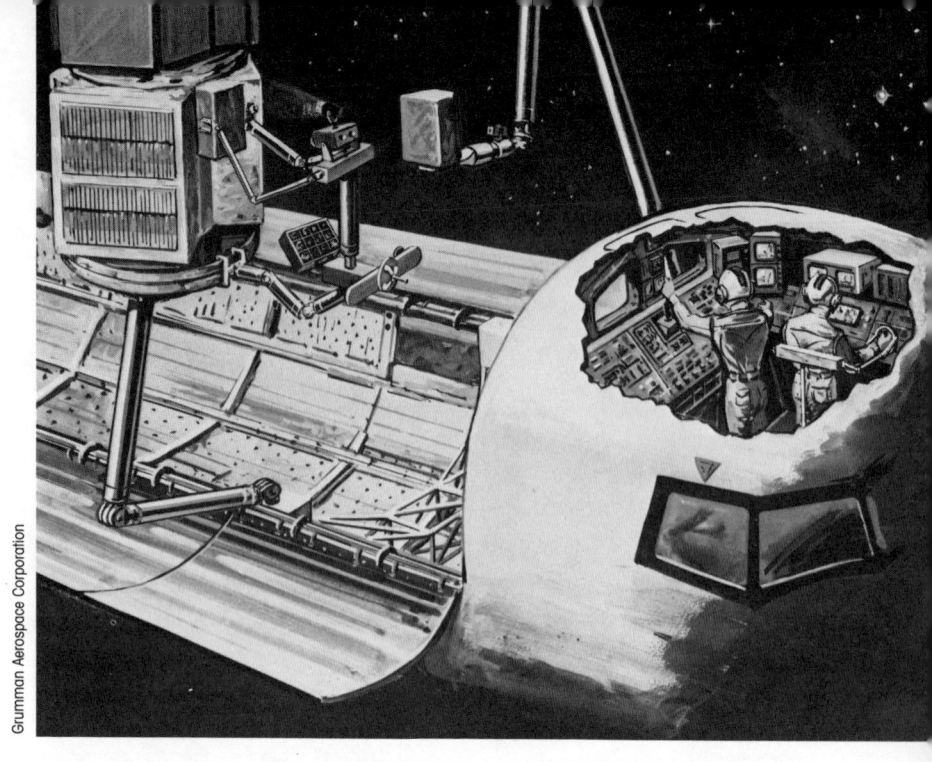

Astronauts flying Space Shuttles will use remote manipulator systems for complicated jobs like repairing satellites.

ideas and Engelberger had vision, drive, and the practical ability to get things going. They found backers and organized a company, Unimation, to develop industrial robots; Engelberger became the president of the company. It took fourteen million dollars of development money and many years before Unimation finally began to show a profit, but now the company is one of the leading industrial robot manufacturers.

One popular Unimation model is the huge Unimate robot used for spray painting, die casting, and other heavy-duty industrial jobs. Unimation's PUMA (short for Programmable Universal Machine for Assembly) is a "human-scale" manipulator, one of the smallest and lightest industrial robots on the market, small enough to fit right into an assembly line designed for humans. This computer-controlled manipulator can be used for very precise and delicate tasks. It can screw light bulbs into sockets, weld metal seams, assemble appliances like electric toasters, and package finished products. A PUMA robot is being used in a candy factory in England to pick up finished pieces of chocolate candy and place them in boxes at a rate of two a second. Human workers tire very rapidly at such a pace, and workers often quit. (None of them had lasted more than two years on the packing job.) A robot packer doesn't get tired or bored, and the bosses don't have to worry about its sampling the chocolates, either.

Cincinnati Milacron, another big robot manufacturer, has a robot called T³, "The Tomorrow Tool." The T³ is very easy to program with a hand-held teaching pendant. It can work on objects while they are moving along an assembly line, and it can do several different jobs in sequence, rather than just repeating a single job over and over.

For a long time Unimation and Cincinnati Milacron were the two companies dominating U.S. robot manufacturing, but now some of the giants of industry are moving into the robot field. In 1982 IBM announced two new industrial robots for use in such jobs as handling parts, drilling, assembling and inspecting parts, packing them into cases or removing them, and machine loading and unloading. One IBM ad reads: "This story begins with the period at the end of this sentence. The robotic arm

The Unimate was designed for heavy-duty industrial jobs.   UNIMATION® Inc.

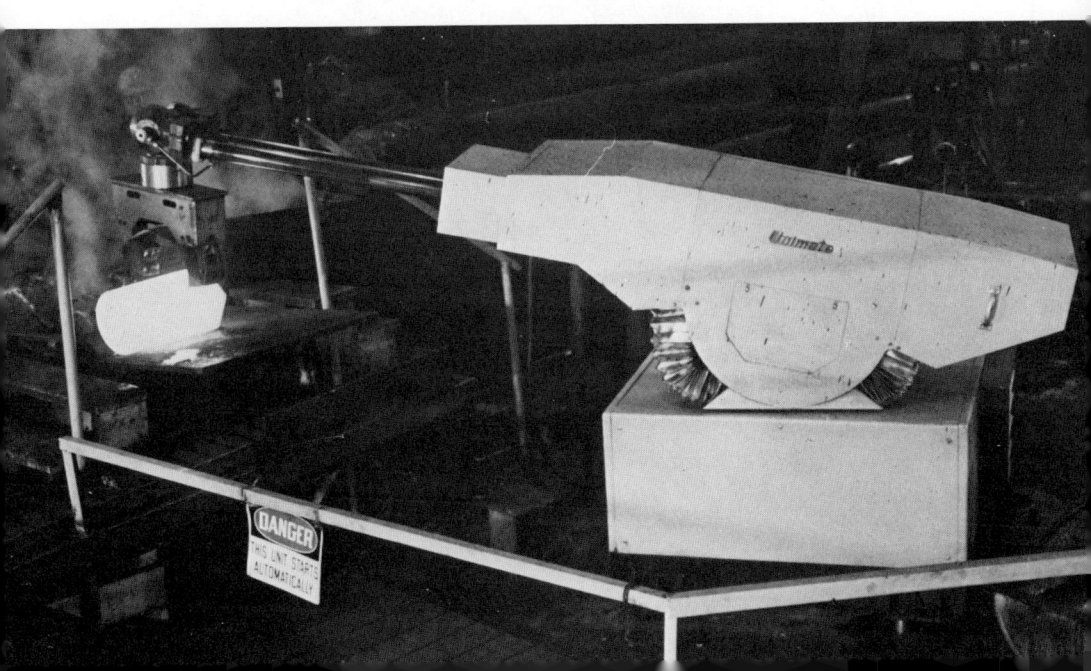

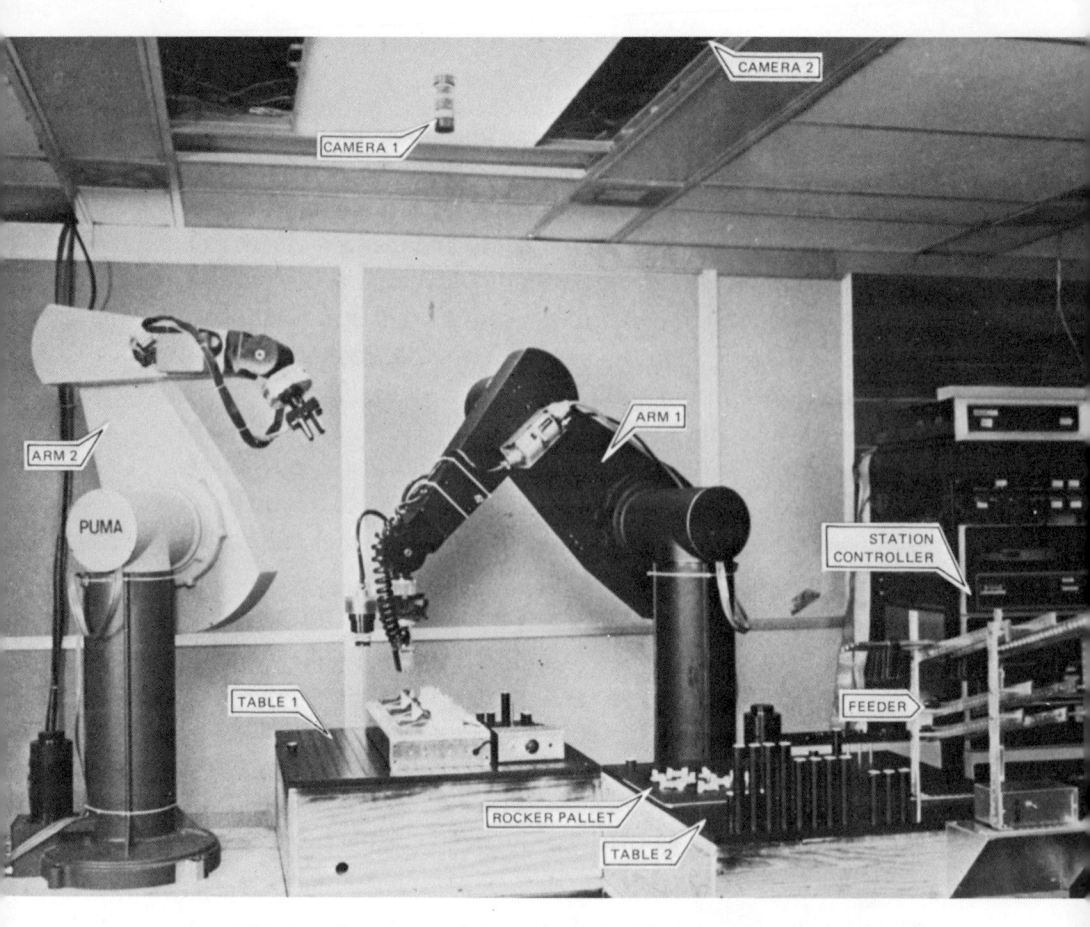

The PUMA is a "human-scale" manipulator. The programmable assembly station shown here was designed at Stanford Research Institute. It includes two PUMA manipulators, a vision system, a click detector, and an assembly table. One PUMA has a camera, while the other has a wrist force sensor.   Courtesy of Robotics Department, SRI International

above can locate a hole that size and accurately insert a pin, once or a thousand times." One of the new IBM robots is designed to work with the IBM personal computer.

Other recent entries in the industrial robot field are Bendix, United Technologies Corporation, General Motors, and Texas Instruments. General Electric has agreed to build and sell robots in the United States for the Japanese firm Hitachi. Meanwhile, new companies are entering the field, such as United States

Robots with a robot called the Maker. (The company is named after the fictional company in the Asimov robot stories.)

Roboticists classify industrial robots in two main groups: non-servo and servo-controlled. (The term "servo" comes from servomechanism, an automatic feedback system for controlling mechanical motion by continual small adjustments, based on information on the position and speed of the parts involved.) Ken Susnjara explains the difference with the example of an electric toy train. One way to stop the train would be to put a heavy block on the track. When the train hits the block, it will stop in exactly the right place. This is a non-servo-controlled system. Another way to stop the train is to watch as it is nearing the desired stopping point, and then to use the electric controls to slow and finally stop it. If you install a sensor that can tell how far the train is from its stopping place and let the sensor operate

With a three-roll wrist, the Cincinnati Milacron T^3 industrial robot can handle tools and make smooth, controlled movements in all directions.

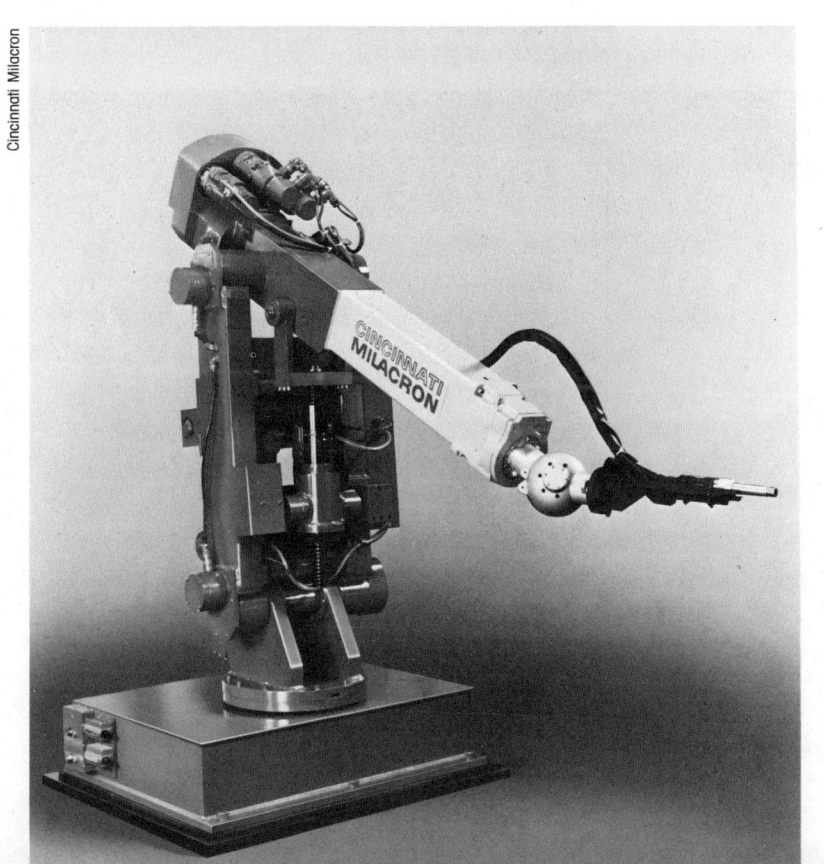

the electric control automatically, you have a servo-controlled system.

Non-servo-type robots are equipped with mechanical stops at the end points of the path. (If stopping points are needed partway along the path, additional mechanical stops can be temporarily inserted and then removed.) The non-servo robots are the simplest and cheapest types. Some of them work electrically, others by means of air valves. (An air valve system has the advantage that it can be used in an explosive atmosphere, where electrical controls might produce sparks and blow up the machine and the factory along with it.) Electrical sequencing controls send signals to the robot to start it on its programmed path. When a signal is received from the robot that some event has happened—for example, that the arm is extended or a clamp has closed—the controller sends a new set of signals for the next part of the program.

The most common type of non-servo robot is the "pick and

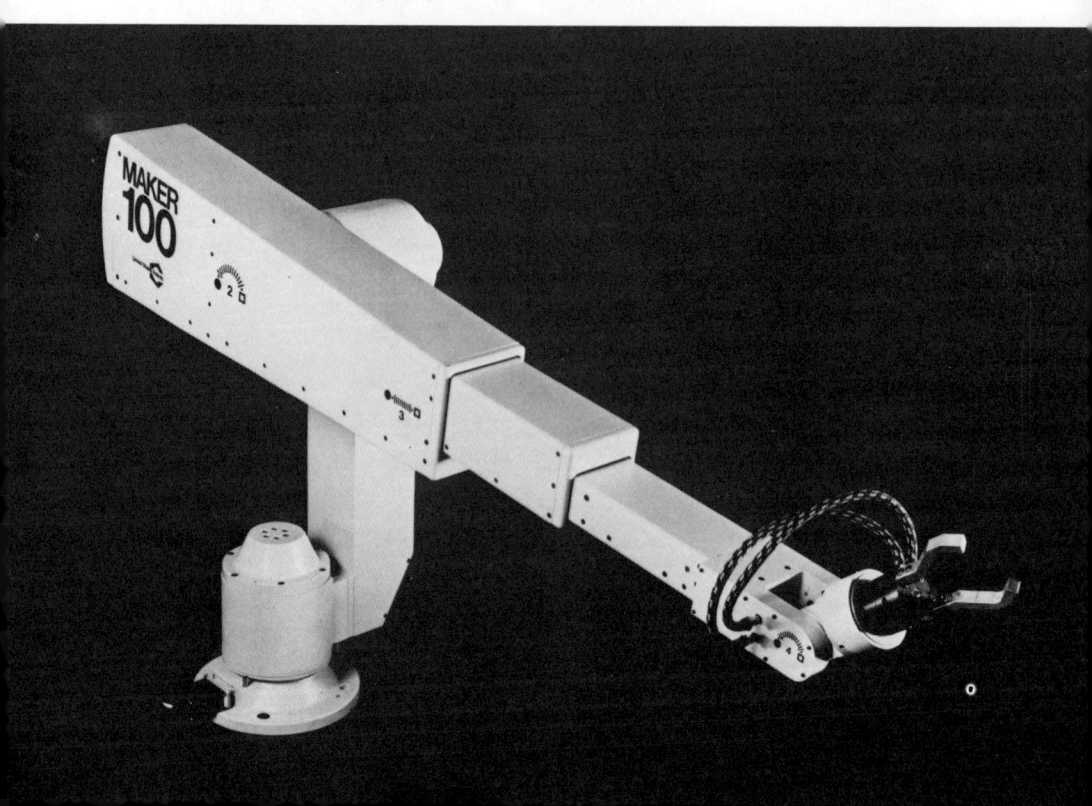

United States Robots' "The Maker" is an industrial robot, not an android like the products of its fictional namesake. United States Robots

place robot." This means that it moves to a position, grasps a part, picks it up, moves to a second position, and puts it into place there. Its freedom of movement is usually limited to two or three directions: in and out, left and right, and up and down. If a pick and place robot picked up a glass of water, it could move it from one table to another, but it could not pour the water out into another glass. The cheapest industrial pick and place robots cost about $5000 in 1983.

Servo-controlled robots are computer controlled and can change directions in midair, without having to trip a mechanical switch. They are capable of a much wider range of movements than non-servo robots, usually in five to seven directions, depending on how they are jointed. Rotation of the "wrist" of a servo robot, for example, permits it to tip a glass over and pour out the water. The robot consists of three main parts: the controller (its computer brain), the manipulator (the robot's base and arm), and the tooling (the hand or gripper). The most sophisticated servo robots, with computer brains and built-in artificial senses, can cost up to $150,000 or more.

You might imagine that industrial robots would be useful mainly in huge factories with enormous production lines. But they can also be a help to small businesses, especially now that there are companies that rent robots rather than selling them outright.

Japanese businessman Toshio Iguchi, for example, runs a small manufacturing business in a shed in the back yard of his house near Tokyo. He makes plastic parts that are used in the production of toy watches. The parts used to be produced and packed into shipping boxes by four part-time workers. But the workers were bored and restless. "At eight o'clock in the morning," Iguchi says, "I would not know whether they would turn up or not. That was vexing."

Then he rented a robot, a swiveling metal arm that slides on a rail and repeats the same motions over and over. For $190 a month, a tenth of what he paid his human employees, the robot forms and packs plastic parts tirelessly. At eight in the morning, Iguchi doesn't have to worry about whether or not his work staff will show up. His robot worker has been toiling all night, and there is a neat stack of boxes filled with red, yellow, and green plastic parts waiting for him, ready to send out to his customers.

Iguchi's former workers have gone on to jobs that they enjoy more, in offices and coffee bars. Their former boss is happy with his robot worker but misses having people around all day. He has started playing golf several times a month and deliberately chooses to play with people he doesn't know. "That way I get to meet people," he says.

Manufacturing industries are not the only places where robots can find a job. Robot mail carriers trundle down the halls of office buildings and even ride the elevators, delivering letters and memos along their routes. Sentry robots are serving as night watchmen in factories and office buildings. Human night watchmen are usually rather poorly paid, so this is generally a "moonlighting" job for them. As a result, they may be tired and inefficient. But a robot sentry will always be alert and will never fall asleep on the job. It uses motion sensors to spot the presence of intruders and smoke and heat detectors to warn of fires.

The Burgerworld International chain of fast food restaurants is planning a robot-staffed restaurant in Ontario, Canada, with three serving robots. Customers will order food directly from the cook, using an intercom. When the food is ready, the cook will program a robot to serve it. The robot waiter, which looks something like R2-D2, can carry four trays at once and can serve nine tables in just 72 seconds.

Police departments in the San Francisco Bay area have been testing a robot called Snoopy. The robot, which runs on treaded wheels, carries a remote-controlled TV camera. In 1982 it helped in the arrest of a man who shot at Oakland police officers and then barricaded himself in his liquor store. The police weren't quite sure where the suspect was hiding, so they sent Snoopy in

In Oakland, California, police used a "Snoopy" robot to scout a liquor store where an armed man was hiding.   Terry Schmitt, United Press International

to have a look around. After they located the man, they sent Snoopy back again with a gun, and it shot out the lock on the door. Then, using its TV camera, it guided the human officers to the suspect.

The New York City police department has bought three Remote Mobile Investigation Units for its bomb squad. These remote-controlled vehicles have robotic arms that can pick up suspected bombs without endangering human officers. The robots also come equipped with a shotgun, a water cannon, a TV camera to inspect a suspicious device, and an X-ray camera to photograph the inside of packages. The human members of the bomb squad were unhappy when they first heard about the robots; they were worried about being replaced by machines. But after they had a chance to see the robots in action, they changed their minds. Now they are grateful for these "high-tech" assistants that are making their jobs safer.

In Australia, sheep raisers are testing a robot sheep shearer. In

Japan, robots direct traffic on highways, and snakelike robots are being designed to creep through pipes, inspect them for leaks, and repair any cracks or holes. Deep in the North Sea, oil companies are testing a robot device that can operate valves and repair damaged parts in underwater oil wells at depths up to 5000 feet. (The deepest levels that human divers can work is 900 feet.) Another diving robot, working for the National Geographic Society, photographed sunken nineteenth-century warships under the icy waters of Lake Ontario.

When robots are used, they generally bring down costs and increase the quality of products. Moreover, robots can often do jobs that humans can't or won't do. Why, then, are there so few robots working today, especially in the United States?

In New York City, robots have joined the bomb squad. Here a robot is retrieving a suspicious briefcase. Police thought it contained a bomb, but it turned out to be holding telephone repair tools. Wide World Photos

Robots excel at dirty, dangerous jobs that human workers don't like to do.

There are many reasons. First of all, people often have a resistance to new ideas; factory owners and managers are no exception. Things are working fine the way they are now, they reason. Why should we bring in expensive new machines that may not work any better? They also worry that their workers may be frightened or angry at the prospect of being replaced by robots. Sometimes that is the case, because robots do eliminate some jobs. Resentful workers may try to sabotage the new robot, either directly (for example, by putting chewing gum in its runners) or more subtly (for example, by adjusting the pace of the assembly line so that parts aren't fed in quite right, and the robot ruins them). This kind of fear and resentment is not really necessary if the management handles the introduction of the robots intelligently.

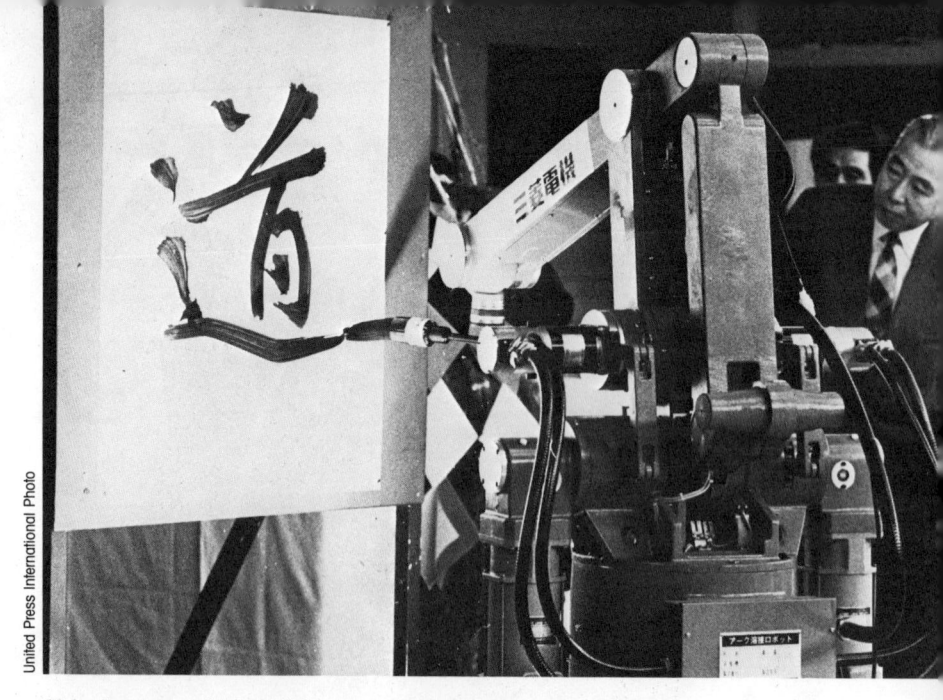

This Japanese industrial robot demonstrates its versatility by painting the Chinese character "michi," which means "road."

There is rarely any need to fire anyone directly after a robot has been installed. There is always a big turnover of workers. People die or retire; they move away or have babies or simply get restless and quit. With this natural loss of workers, it is easy to phase a robot in gradually, simply by not replacing human workers when they leave. During the transition period, the workers whose jobs the robot has taken over can be reassigned or retrained for other tasks.

Since robots do best at the dirty, dangerous, and boring jobs that people don't like to do anyway, the displaced workers may be happier than they were before. And the other workers, knowing that they are in no danger of losing their jobs, can accept the robot and even grow fond of it.

Attitudes are part of the reason for the slow introduction of robots. Economics is another important part. Existing companies have big investments in their current machinery. If they junk those machines and substitute robot systems, they not only have to lay out money for the new equipment, but they have not received full value from the old investments. They may hope that the new systems will be much more efficient and will save them money in the long run, but they can't be sure of that. It can be a difficult decision to make.

In Japan, where robots are far more widely used than in the United States, even with the strictest definition of "robot," the decision was much simpler. World War II destroyed many Japanese factories, and they had to be rebuilt from the ground up. In the process, which took a long time, it made sense to install the most modern and up-to-date machinery available. The development of robots coincided with that rebuilding process. Although Japan is the foremost nation in the world in the use of industrial robots, U.S. robot experts are quick to point out that the basic robot research was done here. The Japanese bought our robots and then adapted them to their own needs.

The economic systems in the two countries have also contributed to a very fast and wide introduction of robots in Japan and a slow introduction in the United States. Here, companies are financed mainly by investments by the public, and stockholders keep an eagle eye on the profits for the latest quarter-year. Investing in major new equipment, such as robot systems, may

This robot wrestler is an electronic game machine in a Tokyo arcade. Its strength can be adjusted by pressing buttons.

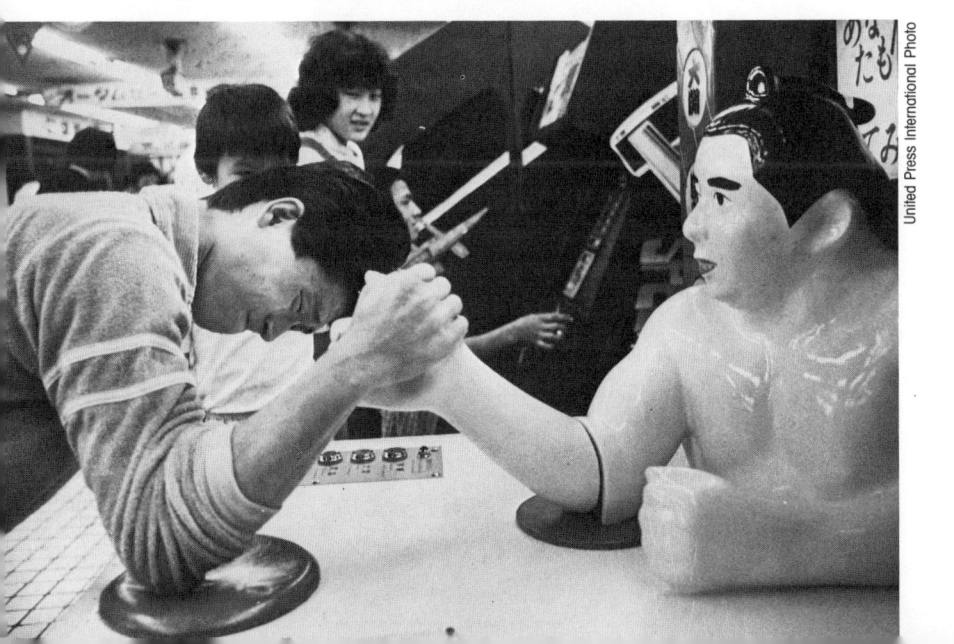

Another industrial robot does tack welding, freeing human workers for less dangerous jobs.

pay off in the long run, but its immediate effect is to cut into profits. So managers of American companies, wanting to be considered successful, tend to plan for the short term. In Japan, on the other hand, companies are financed by banks, not directly by the public. Moreover, the Japanese government tries to help its industries and offers many incentives for improvements such as robots. So Japanese managers can take a longer view. They tend to plan not for three or six months, but instead for five, ten, or even twenty years into the future. With that kind of long-range outlook, robots make a lot of sense.

Another thing that has hampered the introduction of robots into industry is the limitations of today's robots. Ken Overton, a computer scientist at the University of Massachusetts at Amherst, puts it this way: "Consider strapping yourself into a chair with only one arm free. Put blinders on your eyes, plugs in your ears, and a clothespin on your nose. Float in water so you'll feel weightless, put a boxing glove on the one free hand, and try to assemble a mechanical calculator using only chopsticks." With handicaps like that, it's amazing that robots can achieve as much as they do. The development of sensors for robots is helping to take their blinders off and make them more perceptive and responsive to their environment. Meanwhile, though, there is another approach that can also help.

Usually engineers try to fit a robot into work spaces in a factory that has been designed for the needs of humans. Often that works fairly well, but it may not be the most efficient solution. It may be better to design the factory for the needs of the robots. With their limited senses, robots thrive in a predictable environment. The path along which they are programmed to move must be clear; the parts they work on must be delivered at predictable times to predictable places (although some of the most sophisticated robots have more tolerance for variations and errors). In addition, doing things the "human way" may not be the most effective. For example, most robot assemblers insert a screw by a series of half-circle turns, because that is the way a human hand and wrist work. Yet a machine could just as easily be designed to turn continuously until the screw is in place, and that would get the job done much faster.

For a robot worker, the ideal environment would be a completely automated factory, staffed and run by robots and designed for their convenience. Such a robot-run factory has already been built and is operating in Japan. Fittingly enough, it is a factory that builds robots. Parts and tools needed for production are called for by a central control computer, located in the automated warehouse, and delivered to the machine that needs them by a robot delivery cart. One-armed robots pick up the parts and transfer them to the computer-controlled machines that assemble more complex parts, which are sent to the warehouse for storage. (The final assembly stages are still a bit too complicated for robots and are done by humans in another factory.)

During the day shift, the automated factory is staffed by 100 humans, mainly maintenance people who take care of any machine problems that develop. The night shift is rather eerie. In the whole factory building there are only two humans who sit up in a control room and monitor operations on TV screens. Down on the shop floor the lights are dimmed, and the moving machines no longer sound warning beeps as they trundle up and down their dotted lines. Bright lights and warning sounds are not needed, because there are no humans there to see or hear—only the tirelessly working robots.

Perhaps the ultimate in the automated factory of the near future is the CAD/CAM system, which is now a very active area of research and development. CAD/CAM stands for Computer Aided Design/Computer Aided Manufacturing. Engineers, architects, and scientists can use computers to sketch objects in three dimensions. Researchers are working on ways to connect this computer-aided design equipment with computer-controlled machine tools. That way, an engineer can dream up a new design on his or her computer terminal, then press a button that will set machine tools in action—perhaps in a factory hundreds of miles away—to build a working model of the new invention.

Medical Robots

Harvey had several heart attacks yesterday. He'll probably have a few more heart attacks today, and he suffers from nineteen other heart conditions, too. The doctors aren't worried about him, though. Harvey isn't an ordinary patient: he's a Cardiology Patient Simulator—a robot who teaches medical students and doctors how to recognize symptoms of heart disease.

Harvey is the creation of Francis B. Messmore, the president of a New York animation and simulation firm that makes mechanical creatures for movies, plays, and TV commercials. (One of Harvey's "cousins" is an eight-foot-tall rotating Pepsi Free can.) The idea for Harvey dates back to 1969, when a medical school professor asked Messmore to build him a dummy with a heartbeat to use in training medical students. That assignment worked out so well that the professor was soon back with a request for another dummy to demonstrate a heart condition. By the time he had built three dummies, each showing symptoms of a different heart condition, Messmore decided that a single general-purpose heart patient robot was what was needed.

Computer controls inside the cabinet on which Harvey lies

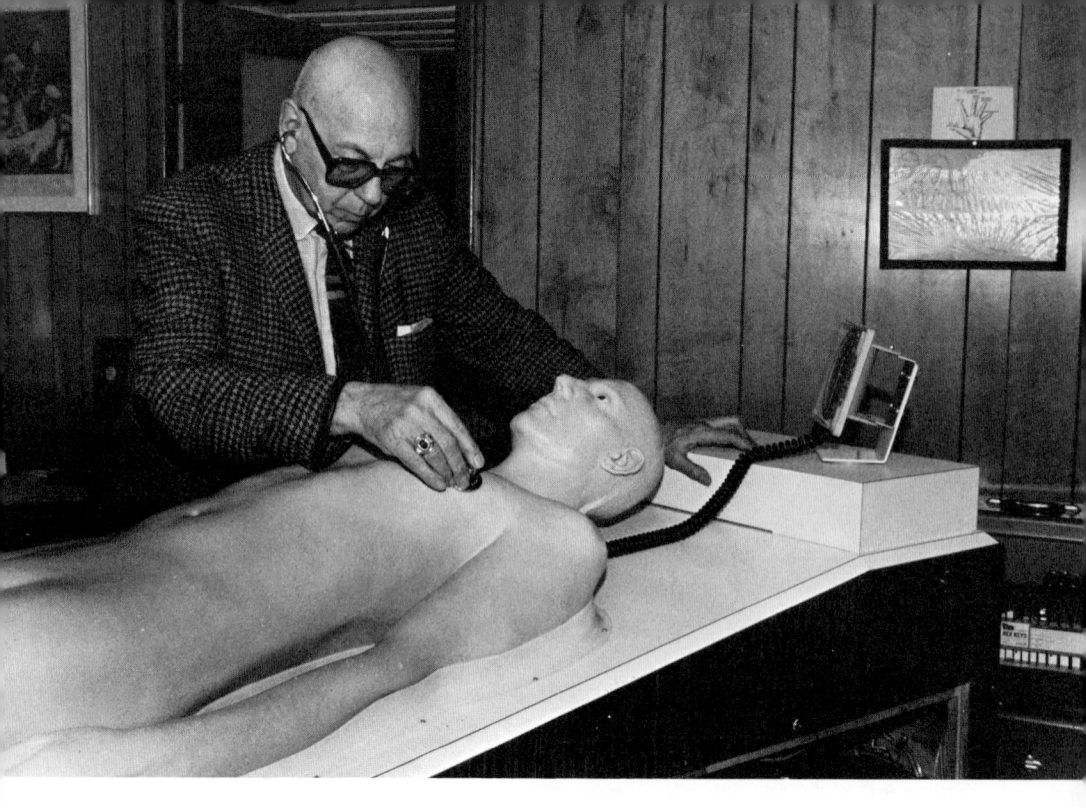

Harvey, the Cardiology Patient Simulator, is helping to train doctors. He is shown below without his skin. Some of the many switches that simulate the symptoms of heart diseases can be seen in the cabinet beneath him.

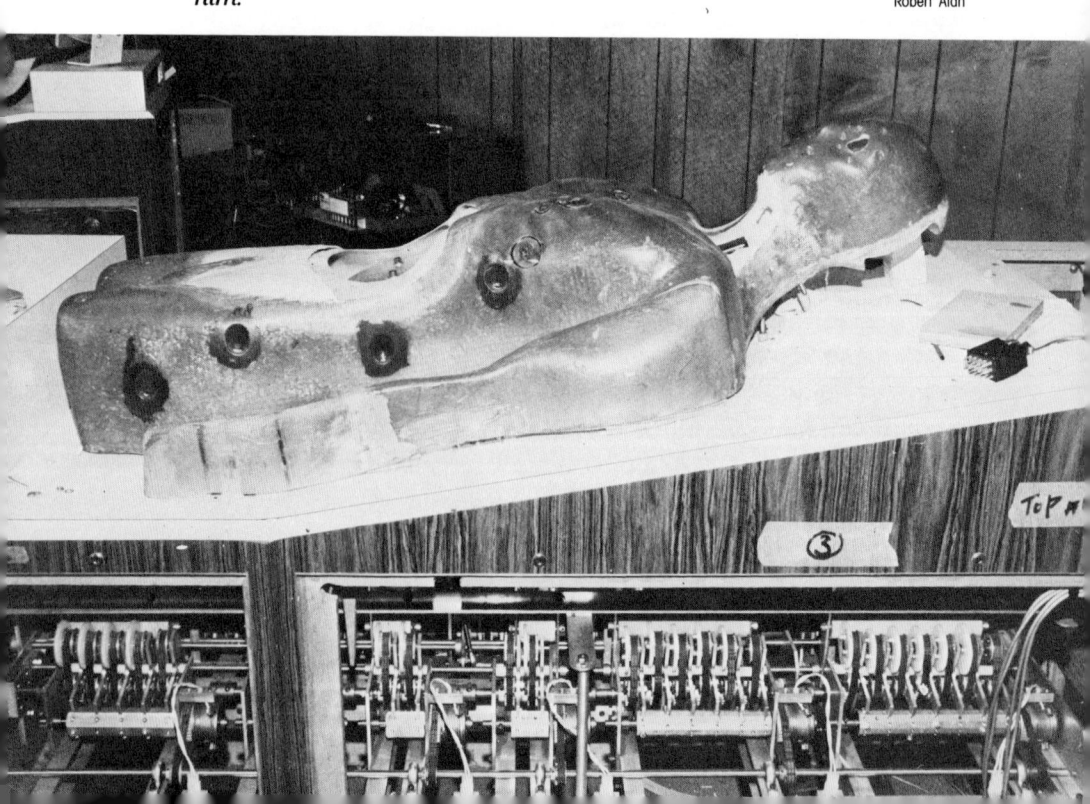

Robert Alan

command more than 400 switches that produce sounds and move mechanical devices. According to the program tape, a touch of a button can make Harvey's veins bulge to show high blood pressure or collapse to signal a sudden pressure drop. A student bending over Harvey's chest with a stethoscope may hear the telltale sloshing sounds that signal a hole in one of the heart partitions or the thump-bump of a heart murmur due to a faulty valve. Harvey's breathing may be fast or slow, regular or irregular, deep or shallow, depending on the symptoms of the disease he is programmed for at the moment.

With a price tag of about $150,000, Harvey is an expensive heart patient. But medical schools all around the world are finding that he is well worth his keep. In studies conducted by the National Heart, Lung, and Blood Institute, fourth-year medical students who had learned on Harvey did twice as well, both on written tests and in examining real patients, as students who studied heart disease the ordinary way. Students can study Harvey at any hour of the day or night, as many times as they want. They never have to worry about waking him or bothering him or perhaps even making his heart condition worse, as they would with a human heart patient.

Other robot patients are also helping to train medical workers. Medical students at the University of Southern California can practice giving anesthesia to Sim One, a lifelike robot with soft plastic skin. Sim One has a pulse and a heartbeat; he turns blue and goes into cardiac arrest if the student anesthetist makes a mistake. Resusci-Annie is another robot teaching tool; she helps people learn to give artificial respiration and CPR.

Robots aren't active only on the patient end of medicine. Researchers are working on the development of robot doctors, too. One medical job that robots can do very well is taking medical histories. Most people seem to enjoy sitting down at a computer screen and answering questions about themselves and their problems. Often they will tell the computer personal details that they would be much too embarrassed to tell a human doctor or nurse. The computer programs are designed to expand on problem areas, presenting additional questions to get a clearer and more complete picture of the patient's symptoms. The computer can organize the information into meaningful patterns that the doctor can use to work out a diagnosis.

Another type of computer program can act as a medical consultant that advises the human doctor on possible diagnoses. One program of this type, called Mycin, was developed at Stanford University to diagnose infectious diseases. Mycin chooses a possible diagnosis from a set of symptoms, making a sort of "educated guess," and then it compares the patient's other symptoms with the known characteristics of the disease to see if they match. If they don't, Mycin tries another diagnosis. At each step of the way, it can be asked to print out its reasoning and the medical references for the information it used. If Mycin makes a mistake in diagnosis, new rules and information can be added to help it avoid the same kind of mistake in the future. In one study, experienced doctors agreed with Mycin's diagnoses 85% of the time.

Other medical consultant programs now being tested include Puff, which helps to diagnose lung problems, and Caduceus, a more ambitious program developed at the University of Pittsburgh to diagnose a variety of diseases. (It can currently handle 600 diseases.) Caduceus asks for more information as it needs it and gradually works out a solution. Dr. John D. Myers, one of Caduceus's developers, says that a computer program like this has some advantages over a human diagnostician. The human memory is just not big enough to deal with all of today's medical knowledge, he says, and the human mind can't juggle so many different possibilities at the same time. The computer can act as a sort of crutch for the doctor.

Today's "robot doctors" are just computers, but it is likely that the medical robots of the future will be much more. Already, electrocardiograms (the recordings of the heart's electrical activity) can be automatically recorded, transmitted, and analyzed. Cancer researchers at Waseda University in Japan are developing a 25-fingered device for detecting breast cancer. Each of the fingers is equipped with a sensitive strain gauge hooked up to a computer. As the fingers examine a breast, the differences in strain allow the computer to draw a map of any lumps that are present. Automated chemical analyzers can test for dozens of chemicals in a drop of blood; sensitive gas chromatographs can analyze the chemicals in a person's breath. Perhaps some day soon devices such as these will be combined into a sort of robot "diagnostic cabinet" that will gather information about the patient's condition and analyze the results to come up with a possible diagnosis.

Meanwhile, a robot at San Diego Memorial Hospital is being trained to assist in brain surgery. The robot is mounted behind the patient's head, and CAT scans (detailed X-ray pictures of a "slice" of tissue) are taken through it. Before, a surgeon would have had to spend hours analyzing the CAT scans to figure out the precise location of a tumor. Then a hole would be drilled through the patient's skull and a radioisotope solution would be dripped into the tumor area. With the surgical robot, the scans are analyzed rapidly by a computer and translated into movements of the robot arm. Within minutes, the robot arm is positioned exactly in the right spot, pointing the way for the human surgeon. Such an operation used to take five to six hours; with robot help it can be cut to only 30 to 40 minutes. When asked how such a robot could be trained and tested, since it couldn't practice on human patients, Unimation President Joseph Engelberger explained laughingly, "Well, they've found all the seeds in a watermelon."

Spinoffs from robotics research are helping to make life easier for handicapped people, just as research on artificial limbs and other prosthetic devices is finding useful applications in robotics.

Damage to the head or spine can leave a person paralyzed, unable to use the legs or even unable to move anything below the neck. Accidents or diseases can also leave a person blind or deaf or unable to speak. More than 35 million Americans today suffer from some sort of physical or mental handicap. Medical science can keep these people alive, but caring for them can be very expensive, and without the ability to see or move or speak, their lives may be sadly wasted. Robotic devices are being developed to help handicapped people care for themselves, work, and live a more normal life.

One early approach to helping people suffering from paralysis was to build mechanical devices that could move their arms for them. But these contraptions were usually heavy, bulky, and very limited. In addition, there was the danger of accidentally pulling the person's arm too far and injuring it. (A paralyzed person doesn't receive the usual pain warnings of danger.) Researchers decided that what they needed was a device *like* an arm to substitute for the arms the handicapped person could no longer use. In the mid-1970s, robotics researchers from the NASA space program, working at the Jet Propulsion Laboratory in Pasadena, developed a voice-controlled wheelchair equipped with an electromechanical arm for quadriplegics (people who had lost the use of all their limbs). The wheelchair recognized 36 commands, which directed it to move around and pick up things for the person sitting in it.

A more sophisticated manipulating system for quadriplegics has recently been developed by Stanford University researchers working at the Veterans Administration Center in Palo Alto, California. It uses a robot manipulator designed for industry, the Unimation PUMA-250, which has special sensors to find objects and grasp them precisely and securely. (It can move a heavy weight or pick up a delicate wineglass without breaking it.) This Robotic Aid has a microcomputer brain that can translate voice commands into movements of the robot arm, an additional set of simple "joystick" controls that a paralyzed person can operate with his or her head, and a voice synthesizer that permits it to "talk back" to its user. The user can program the Robotic Aid for an individual set of tasks. First the handicapped person breaks down the tasks into a series of movements for which the robot

Robots are helping the handicapped. The robotic arm and worktable (above) were designed for paralyzed patients. Melkong, the robot hospital aid (below), is being developed in Japan.

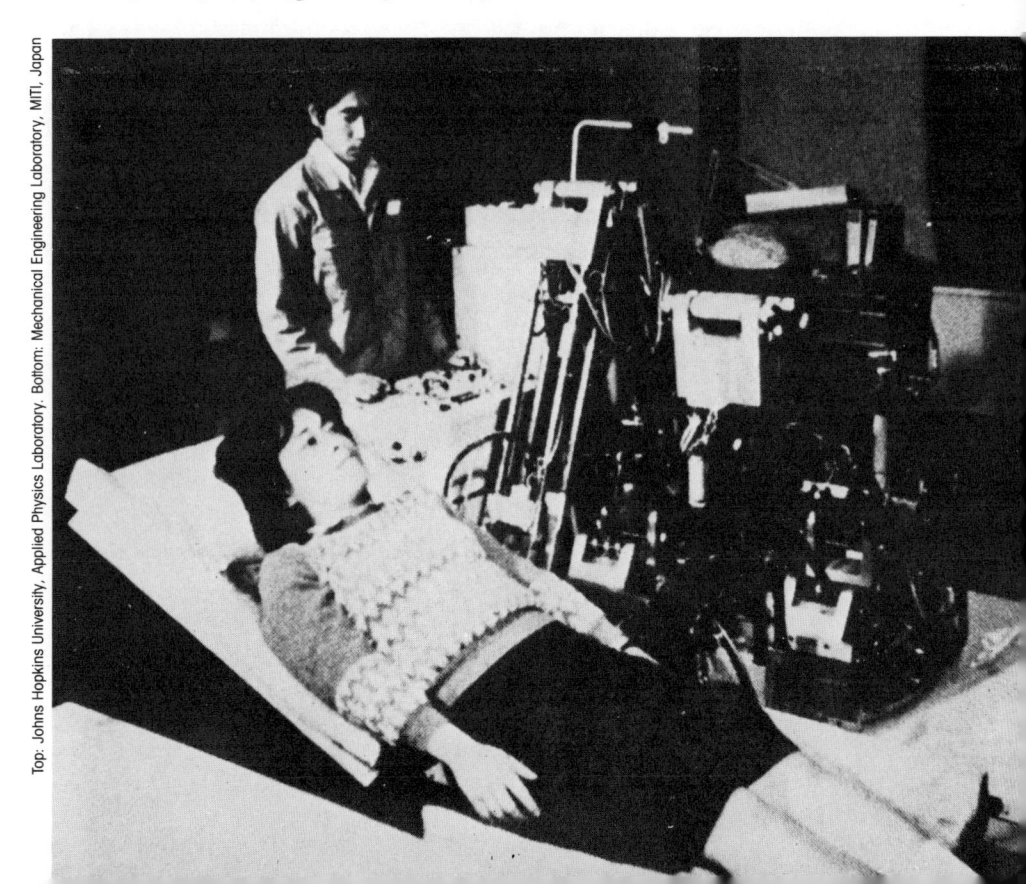

Top: Johns Hopkins University, Applied Physics Laboratory. Bottom: Mechanical Engineering Laboratory, MITI, Japan

already knows the spoken commands. The robot stores the series of commands in its memory, and when the task is finished, the person gives it a name. Now all he or she has to do is to say the name of the task, and the robot will perform all the movements it requires. The Robotic Aid responds only to the voice of its "master," but it can learn to work for someone else if it is given some new instructions. The device can be used to fetch books and other objects, to help the person eat, and to do other useful tasks.

The Palo Alto researchers are now working on an improved Robotic Aid mounted on a mobile base and equipped with a vision system. The handicapped person could program the robot to find its way around the house, to bring back a book from the bookshelf or a snack from the refrigerator.

Meanwhile, researchers at Japan's Medical Precision Engineering Institute are working along similar lines. They have developed a pair of manipulator arms mounted on a robot cart that moves back and forth between a patient's bed and a storage cabinet. The patient controls the robot by means of a keyboard, voice commands, or even by whistles and gasps. The robot can bring the patient any object that is stored in the cabinet, such as a newspaper or a piece of fruit. The manipulator isn't quite deft enough to peel the fruit yet, but the researchers are working on that, too. Another Japanese robot hospital aid is called Melkong (for "mechanical electrical King Kong"). Melkong moves patients—very gently—out of bed, into a bathtub, and back into bed again.

The same kind of vision systems that are being developed for robots are also being used in experimental systems to bring mobility and sight to the blind. Today's personal robots use ultrasonic scanners or optical sensors to help them move around without bumping into things. Blind people can use devices like the Sonicguide, which bounce beams of ultrasound or light off objects and convert the reflections to warning sounds or vibrations. Robot vision systems typically build up a crude

The MELDOG is a robot "seeing eye dog" developed in Japan.

two-dimensional picture of objects from the contrasting patterns of light and darkness. Vision systems for the blind use a miniature TV camera to build up a similar picture.

In one system, under development at Smith-Kettlewell Institute of Visual Sciences in San Francisco, the picture is translated into a pattern of dots drawn by vibrating cones on the skin of a person's belly. It takes time for a person to learn to "see" images from skin sensations like these, but blind students have learned to do it well enough to find objects around a room, read meters, and even use scientific instruments.

In another approach, researchers at the University of Utah and the Institute for Artificial Organs in New York implant electrodes into the vision center of the blind person's brain. A computer translates the pictures from a TV camera into points on the

The Sonicguide is a navigation system for the blind.

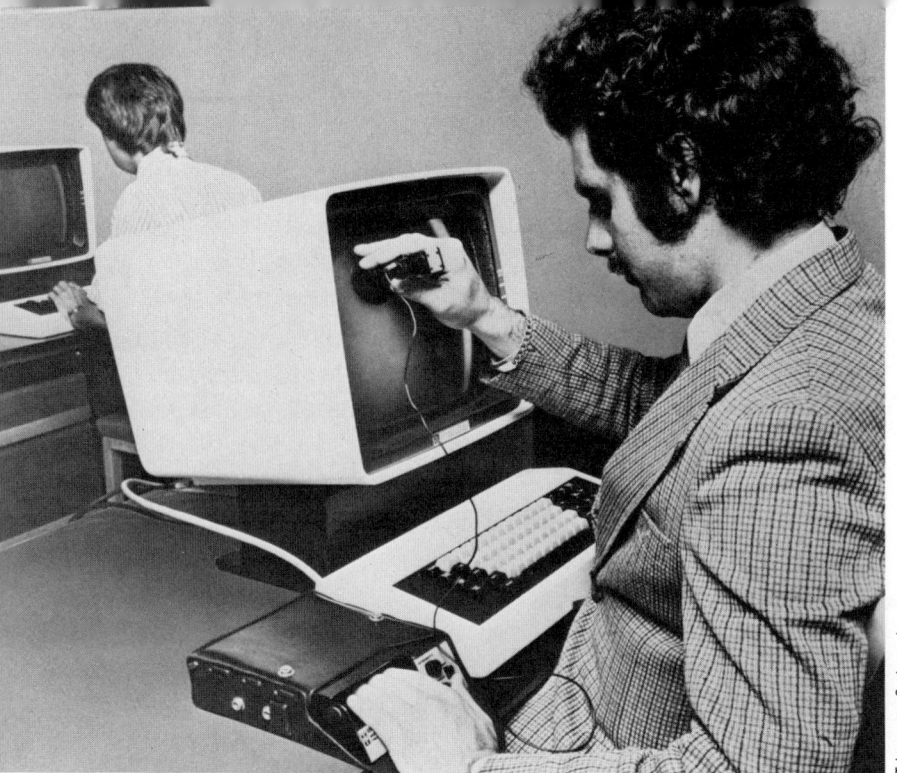

The Optacon robot reading machine scans a page of print or a computer display screen and translates the letters into vibrations felt with the fingertips.

person's own "vision map," stimulates the appropriate electrodes, and the person "sees."

Learning to use one of these new vision devices is very difficult, and the systems are still quite experimental. But many blind people are already benefitting from robotlike machines for reading print. The Optacon scans pages of print or computer displays and uses a computer to convert the letters into vibrations that the user can feel through the fingertips. The Kurzweil machine goes one step further, using voice synthesizers to read the text out loud. (Voice synthesizers are also bringing a more normal life to people who have lost the power of speech.)

As robotics research and robot applications expand, the new developments will also help bring better artificial organs, improved "smart machines," and other helpful new tools for medicine.

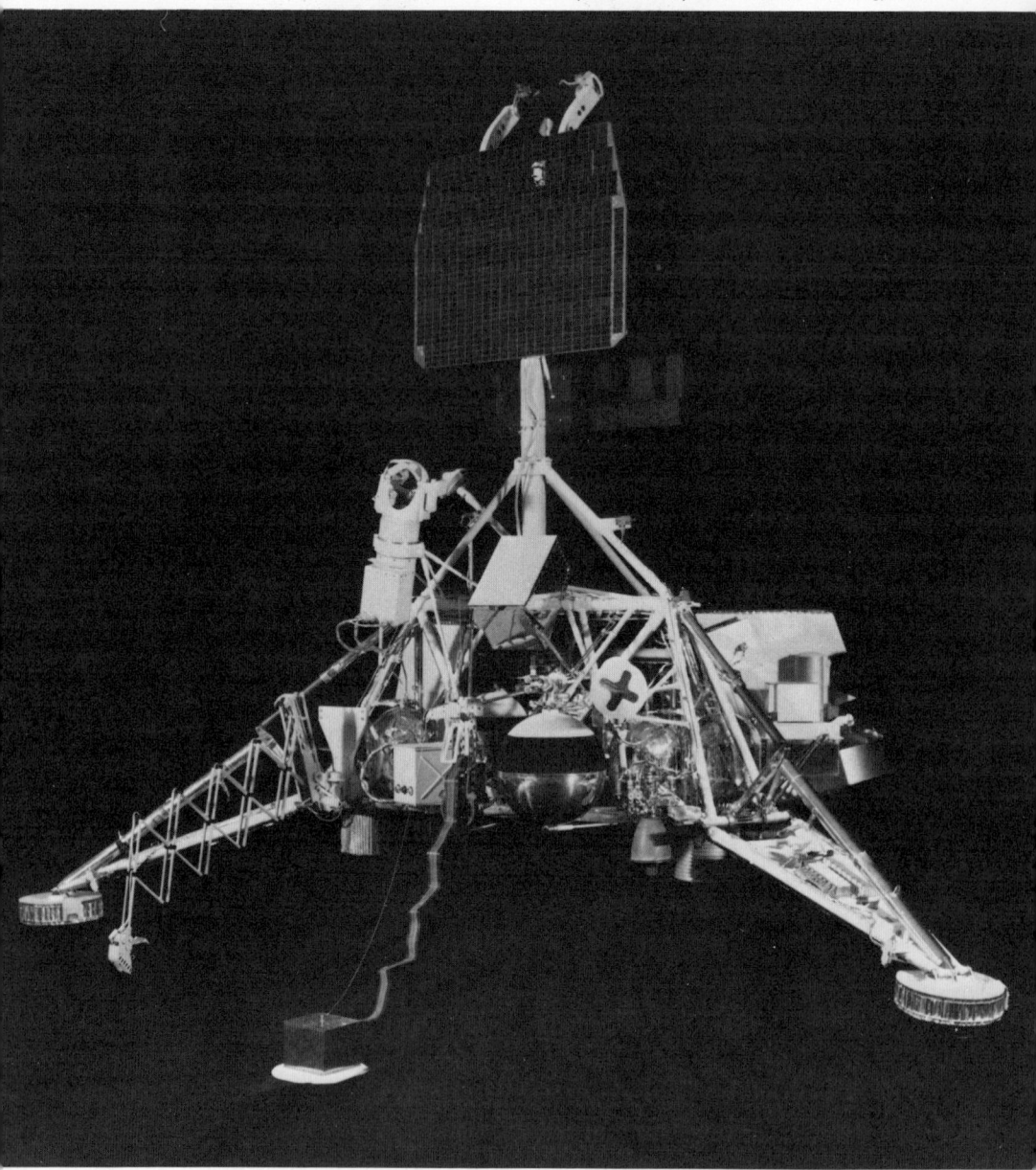

Jet Propulsion Laboratory, California Institute of Technology, NASA

Surveyor 7 landed on the moon in 1968 and returned TV pictures and data on the composition of lunar soil.

Robots in Space

If you think about our exploration of space, chances are that the first thing to come to your mind will be astronauts—humans manning the Space Shuttle, bulky spacesuited figures floating weightless in a "spacewalk," and the highlight of our ten-year crash program to put a man on the moon: Neil Armstrong carefully lowering his foot onto the lunar soil and saying solemnly, "One small step for man. . ." Public interest and excitement was at a peak during the Apollo flights, but then, when the emphasis in the space program shifted back to unmanned flights, people lost interest.

It's a pity that unmanned space flights don't have as much popular appeal as those piloted by astronauts. The robots that we have already sent out into space and those that are planned for the future represent an amazing engineering achievement. In many ways, they are the most practical way to explore space and develop the resources of our neighboring moons and planets.

Robot spacecraft prepared the way for the first successful manned landing on the moon in 1969. Rangers 7, 8, and 9 had an automatic system to turn their cameras on 13 minutes before the spacecraft was due to crash-land on the moon. They sent back the first close-range photographs of the lunar surface. Then came the Surveyor, which made a fully controlled soft landing on the moon in 1966. Sophisticated scientific equipment on the Surveyor was remote-controlled by signals from humans on Earth. Only after these robot spacecraft had accomplished their tasks were men sent to the moon, a feat that caught and held the public imagination for years. Behind the scenes, though, NASA researchers were still hard at work on automated systems and devices—space robots. They were thinking ahead.

We earthlings have long been curious about the neighboring planets that we can see shining in the night sky. Fiction writers dreamed of exotic creatures with advanced civilizations living on Mars and Venus. Our best telescopes on Earth couldn't tell us whether or not those dreams might be true. Venus was hidden in a cloak of thick clouds. Mars seemed to have a bit of an atmosphere, along with some intriguing surface markings that changed with the seasons and were called "canals." Then came the spacecraft that sent back more information about our neighbors. Venus turned out to be incredibly hot; there isn't likely to be any life there, at least on the surface of the planet. Mars was still a possibility, though. But spacecraft on flyby missions and even those that crash-landed could send back only a limited amount of information. They couldn't give us *real* close-up pictures fine enough to spot a living Martian.

What was needed was a mission to Mars. But it couldn't be a manned mission, even if we had another crash program like our man-on-the-moon effort. It is simply too far. We're not ready yet for a manned Mars mission; we won't spend the enormous amounts of money necessary to develop systems to support a human crew on that long a trip. So NASA decided to use robots. They would have to be much more sophisticated than the Surveyor landers that sent back information on the moon. They couldn't depend on remote control from Earth, the way the Surveyor did.

Why not?

It takes time for a radio message to travel through space. And the farther the message has to travel, the longer the time lag. The Earth-based controllers of the Surveyor moon landers used a "move and wait" method of control. They would send a message for a particular movement and then wait for the return message to tell them what the result was. Then they could send a message for the next movement. Each message to the moon takes two and a half seconds for a round trip. That doesn't sound like very long, but if you try lifting your foot and then counting off two and a half seconds by the clock before you put it down, you will understand that even that time lag could become irksome. It is workable, though.

Scientists in the U.S. were able to manipulate their scientific instruments on the Surveyor landers, and in 1970 Russian scientists used radio messages to control a little robot lunar rover called the Lunakhod. This vehicle, which was about the size of a queen-sized bed, ran on electric motors and solar-powered batteries. Scientists on Earth controlled its movements and monitored them on television viewers, but the Lunakhod could make some on-the-spot decisions of its own. It could sense when it ran over a large rock and started to tip. When that happened, it would automatically command itself to stop and wait till its human masters gave it a new command to correct the tilt. If one of the Lunakhod's eight wheels got stuck in the lunar sand, the Russian controllers had a rather drastic way to solve the problem: they sent a radio signal that burst the driveshaft of the bogged-down wheel, permitting the other wheels to take over. Obviously that was a solution that couldn't be used too often, but before it ran out of working wheels, the Lunakhod had covered six and a half miles over the lunar terrain in 322 days of exploration.

Although remote control worked for moon missions, it was out of the question for robot explorers of the planets. Even the closest planets are many times farther away from us than the moon; radio messages take proportionately longer to travel and return. Earthbound controllers trying to work a Mars rover by radio signals would have to wait up to 40 minutes to find out the effect of each signal! Imagine the situation: the rover is trundling along when it comes to a crater and starts to tip over the edge. Twenty minutes later the television transmission from the rover informs its mission controllers of its plight. Urgently they radio the correction signal: STOP! But it takes the signal 20 more minutes to reach the rover, which is now an expensive scrap heap at the bottom of the crater. Obviously a Mars rover has to be able to make many decisions on its own, without waiting for messages from a controller on Earth.

In 1976, after a voyage of more than 300 days, two Viking spacecraft went into orbit around Mars, photographed possible landing sites, and then sent down robot landers. The Viking landers worked beautifully, making split-second decisions on their own, monitoring the atmosphere and their own positions, opening parachutes and firing retro-rockets at just the right times to make perfect landings. Then they set about making observations and conducting experiments. They measured the Martian atmosphere, wind velocity and direction, listened for "marsquakes," and took samples of soil with a manipulator arm. Biological experiments tested the soil to see if there were any bacteria or other living creatures in it. Some of these actions were pre-programmed; others were remote-controlled from Earth.

After analyzing the results of the tests, scientists concluded that there weren't any signs of life on Mars—at least in the areas where the two Viking landers had set down. But those were only two very small spots on the planet, and they might not tell the whole story. (What kind of results would a Viking lander come up with if it set down in the middle of the Sahara desert here on Earth?) To be surer about the conclusions and to learn more about our neighbor planet we need a Mars rover that can move around to explore a larger area.

A model of the Viking landers that managed their own soft landings on Mars in 1976 and conducted experiments there.

NASA engineers and scientists at the Jet Propulsion Laboratory in Pasadena, California, have been working on a Mars rover that can operate without continual instructions from human controllers. It looks something like a golf cart and moves by means of four loopwheel legs with jointed "knees." Each loopwheel foot can adjust independently to varying heights and depths, while the boxlike chassis still stays level. The Mars rover can thus step on rocks and climb up and down hills without tipping over. It travels at a speed of about one meter per minute. Two TV cameras, working in stereo like human eyes, provide a vision system. In addition, a laser range finder can measure the distance between the rover and any obstacles. The rover's computer brain uses this information to chart the best course and instructs the rover to move along that path. Manipulator arms coordinated by the vision system and pressure sensors in the wrists can be used to collect soil samples and to perform other tasks.

Such a Mars rover could roam around the Mars terrain, observing and relaying pictures back to Earth. Messages from human controllers could send it on a different path or ask it to do particular experiments, but most of the time the rover would be on its own.

Unfortunately, work on the Mars rover has just about stopped for the present, because the Mars mission has been put off for at least ten to fifteen years. But the Jet Propulsion Lab researchers are still working on vision systems, improved sensors, and robotic intelligence, which will help to improve later models of the Mars rover and other space robots.

A variation of the Mars rover that is being developed to travel over the Mars terrain without instructions from human controllers.

Jet Propulsion Laboratory, California Institute of Technology, NASA

Exploring planets is not the only use for robots in space. The Space Shuttle program will soon be leading to an explosion of activity. Huge solar energy collectors out in orbit that will beam power down to Earth in the form of microwaves are on the drawing board. Many industries are interested in the possibilities of manufacturing in space. Some materials have special useful properties under vacuum, at zero gravity, and at ultra-low temperatures. These conditions can be produced on Earth, but only at great expense; out in space they are all free.

Space mining will also be an important industry of the future. Tests of moon soil and rock indicated that valuable minerals could be mined rather easily there. Mining on the asteroids—the rocks that circle the sun in an orbit beyond Mars—also seems promising.

Who will be filling all the mining, manufacturing, and service jobs out in space? The obvious answer might seem to be people. But there are many reasons why that answer may not be the best one. Human workers would have to be transported and brought back. With the huge expense involved in each rocket launching, even with reusable space shuttles, working in space would not be a question of putting in an eight-hour shift and then catching the commuter shuttle back home. Space work crews would have to sign on for long work tours—months at least—which would mean that facilities for eating, sleeping, medical service, and recreation would have to be constructed out in orbit.

A NASA study indicated that it would cost at least $2 million to maintain a single crew member in orbit for a year. And yet, human labor out in space is rather inefficient. A person in space needs a bulky spacesuit with elaborate life-support systems. And working out there is not only dangerous, but also very hard on a body that is adapted to life on a planet with Earth gravity and a breathable atmosphere.

The astronauts' experiences have shown that a person can safely perform only one or two hours of extravehicular activity (an activity performed by an astronant outside a vehicle in space) during each 24-hour period. If a space technician worked a whole year on an orbiting space station without taking any days off for vacations or sick leave, the useful extravehicular activity would cost about $3000 an hour, without even counting the technician's salary!

Robots are a better answer for work out in space. They don't need to breathe air, zero gravity doesn't bother them, and they never get tired. NASA researchers are developing a variety of robots for building and repairing space stations and for working in them. Even on the moon and neighboring planets, it would be much cheaper and more efficient to support a large robot work force with perhaps a few human supervisors, instead of trying to maintain a large human staff.

One of the most intriguing ideas is to design robots that not only mine the rocks for their valuable minerals and manufacture useful products, but also produce more robots like themselves. These robots would form a steadily expanding work force that could "live off the land," taking the materials they need from the surface rocks and soil. Eventually such robots might even be

Manufacturing out in space will use mainly automated machines and robots. This Beam Builder, developed by NASA, can form beams of any length from aluminum sheet at a rate of about three feet per minute.

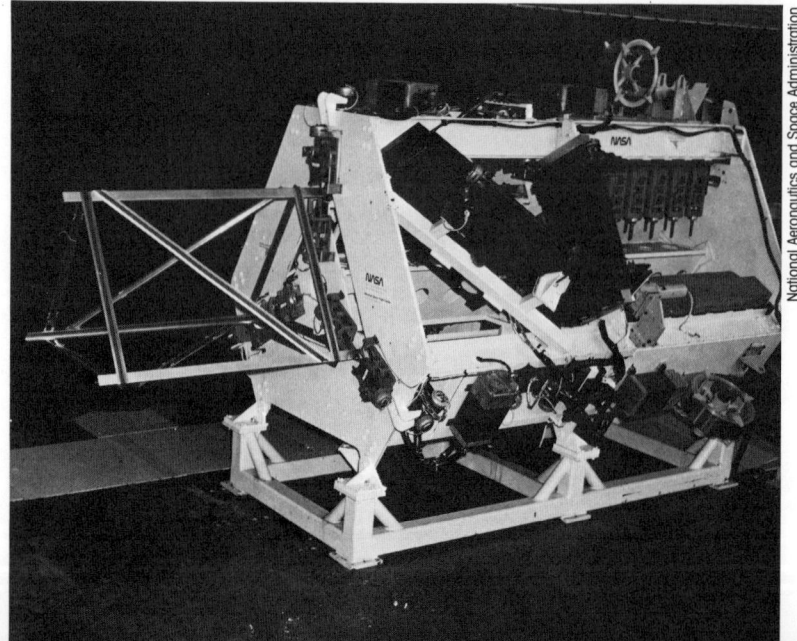

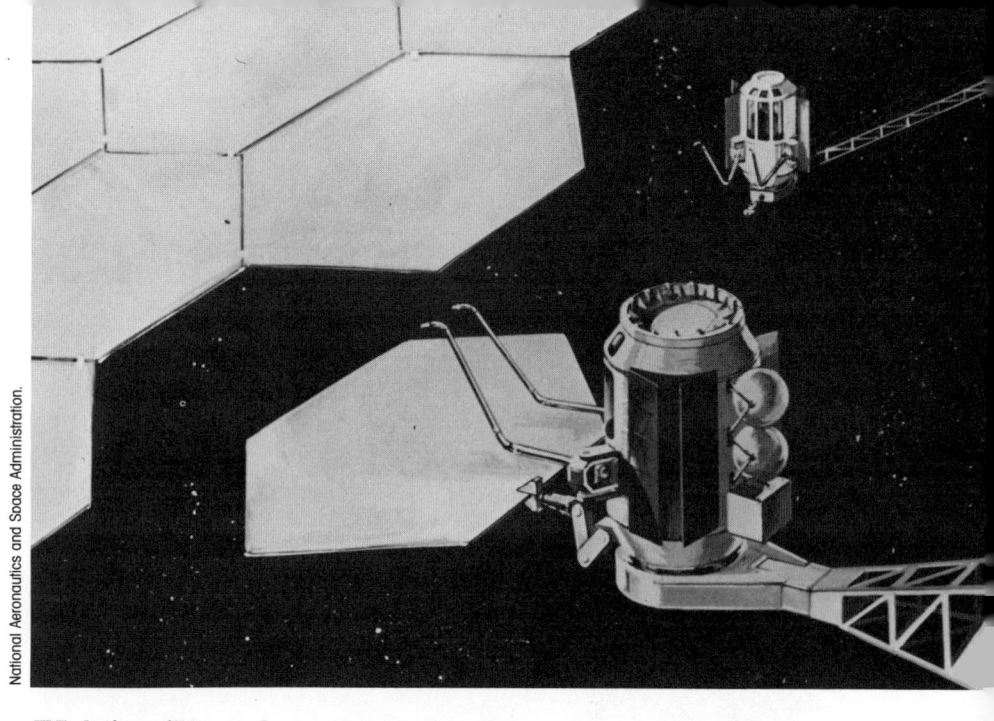

EVA duties will be performed by robots and remote-controlled manipulators. The "cherry picker" design holds a human operator who works with robot manipulator arms on tasks like repairing satellites and building solar collectors.

used to "terraform" the moon and planets such as Mars: Their mining and manufacturing work could be designed to generate oxygen as a by-product, and that could gradually build up to form an atmosphere that humans could breathe.

Self-reproducing robots and terraforming of planets are projects fairly far off in the future. But closer to our time and closer to home, robots are already being developed to work in dangerous and hostile environments on our own planet. The crushing pressures and water environment of the ocean bottom are just as dangerous for humans as the weightless vacuum of space. Semiautomatic robot explorers are already prowling along the ocean bottom, collecting rock samples and sending signals back to their human controllers in the ships above. Concentrated lumps of valuable minerals have been discovered, and plans are being made to mine them. Robots will be the best way to gather these riches from the deep.

Heath Company, Benton Harbor, MI

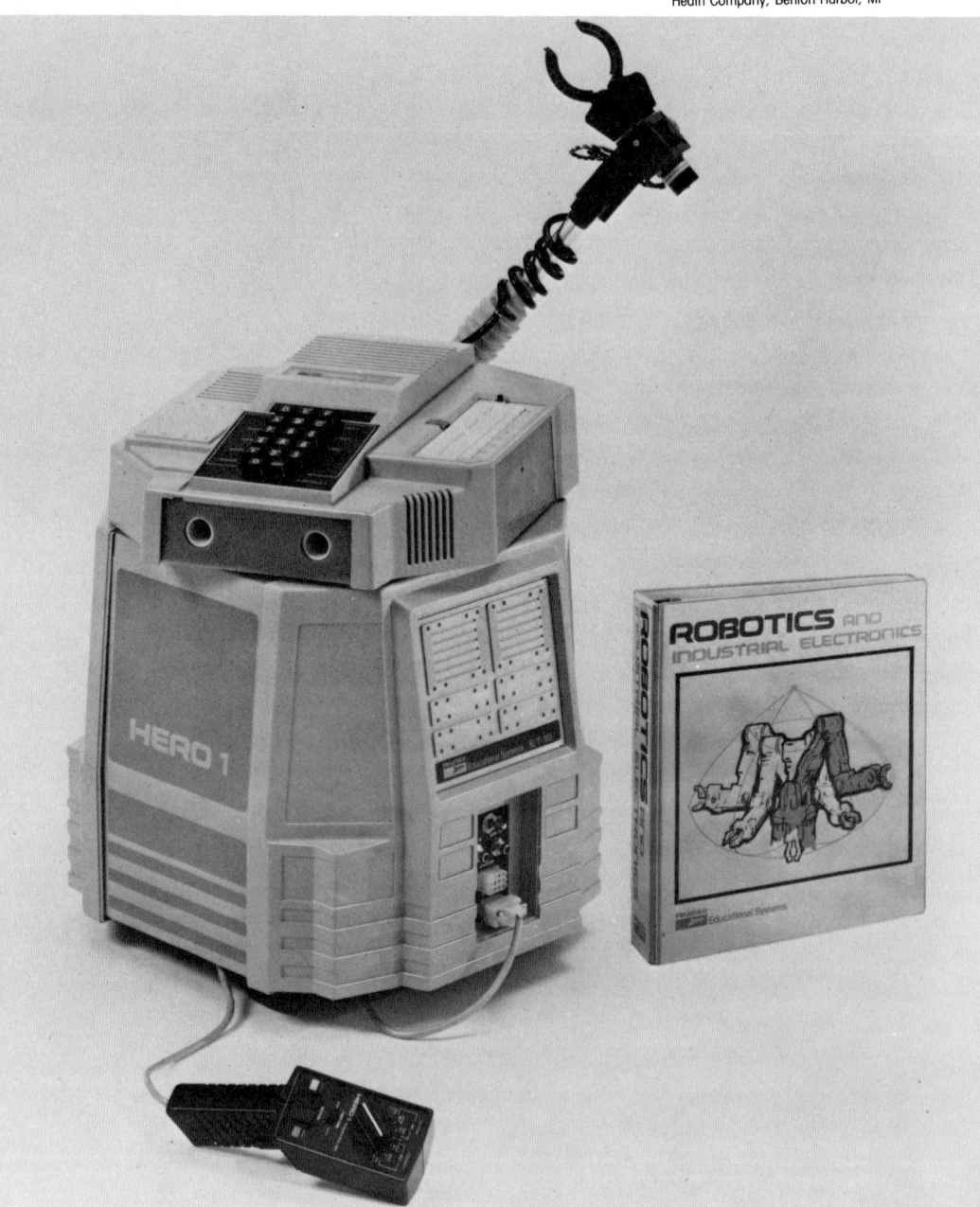

Heath's HERO I, one of the first of the personal robots.

A Robot in the House

Have you ever dreamed about having a little "mechanical person" around the house? A robot who'd clean the floors and wash the dishes; one who'd talk to you, play games with you, maybe even help you with your homework? Your own personal mechanical servant, obedient to all your commands? If that's your fantasy, then your dreams are coming true! The personal robot is here at last. . .well, sort of.

In 1983, robots seemed to be everywhere. Magazines had articles about the "Year of the Robot." At a community college in Maryland, a robot delivered the commencement speech. If you turned on the TV news, you were likely to see a fat little robot rolling around the set, shaking hands, picking things up, and talking in its funny little robot voice. "Home robots" were for sale at last, at a price that many people could afford. Averaging about $1500 to $2500 each, they cost more than a TV set but much less than a car. These first-generation home robots, though, are still far from the efficient mechanical servants described in science fiction stories. They're not as bright as a pet dog, and their abilities are very limited. But they're getting smarter and more capable all the time.

The first personal robot to grab the headlines was the Hero I, announced in December 1982. Hero I is produced by Heath, the makers of the famous Heathkits for electronics hobbyists. Hero I comes in kit form, ready to put together. For people who don't feel up to building their own robot, even with detailed instructions (Heathkits are not for beginners!), the Hero I can also be bought factory-assembled.

Hero I is a squat little creature that looks very much like R2-D2 from the movie *Star Wars*. It stands 20 inches tall and weighs 30 pounds. It has a turretlike head that can turn around through almost a full circle, and it rolls around on a three-wheeled base. Its electronic "eyes" and "ears" can detect light, sound, and motion. Its ultrasonic ranging system bounces ultra-high-pitched sound beams off objects in its path to keep it from bumping into things. Hero carries its own rechargeable power supply and its own on-board programmable computer, as well as the motors that drive its wheels and its arm. (The arm is attached to Hero's head.)

What can Hero do? It can roll around on a preset pathway, pick things up with its arm, and its built-in voice synthesizer permits it to talk and even sing. Hero gets its instructions through programs typed into its on-board keyboard or on a separate "teaching pendant," which looks like a hand calculator. Once it has been taught to do something, it remembers its instructions and can repeat the movements. Hero's programs can be stored on cassette tapes, like those used in home computers. With a little work, Hero's owner can get it to respond to noises and motion. For example, you could teach Hero to say, "I heard that!" when you clap loudly, or to yell, "Warning, Intruder! I have called the police!" if it spots movement in the doorway. Your parents could program Hero to tell you to turn down the stereo if you are playing it too loud.

Hero's name is an abbreviation for Heath Educational Robot. It

Heath Company, Benton Harbor, MI

HERO I was really intended as a teaching tool for robotics workers. Its popularity as a home robot took its makers by surprise.

wasn't really designed as a home companion, servant, or pet. Hero is a teaching robot, meant to be used in training robotics students. It comes with a 1200-page course in robotics and was carefully designed to include all the basic systems found in modern industrial robots. It also has an experimental circuit board that lets the operators add new attachments that they have designed.

As a personal robot, Hero has quite a bit of competition. Another one in the running is the RB5X, produced by RB Robot Corporation in Golden, Colorado. Standing 24 inches tall, RB5X looks a little like a fire hydrant. It has a transparent dome, six

81

The RB5X is an electronic pet.

"feely bumpers" mounted around its base, and two wheels, each powered by its own little motor. RB5X is a self-learning robot that can learn from experience. Its built-in computer is programmed for one main motivation: to keep moving. At first RB5X moves randomly, rolling along until it bumps into something, which causes it to turn to the right or left. It remembers which of its six touch sensors was bumped, and it also takes note of whether a particular turn—a right turn, for example—allowed it to keep moving. Gradually, moving and bumping, it builds up a map of its environment and learns to follow a path that will permit it to keep moving without getting bumped. (Later the operator can clear RB5X's memory to let it learn a new path, which may be quite different but just as successful.)

In addition to its touch sensors, RB5X has a sonar ranging system and a photocell mounted on its undercarriage. That might seem like a rather odd place to mount a robot's "eye," but RB5X's photocell is very useful: when the robot is "hungry," it can follow a white tape to a recharger and plug itself in. (It can go for about eight or nine hours on a battery charge before it gets "tired.")

The RB people plan lots of add-on options for their robot, such as an arm to pick up and move things and even run a vacuum cleaner. Gradually the options will be incorporated into the standard model, as the robot graduates from RB5X (the "X" stands for experimental) to RB5 and then to RB4, RB3, RB2, and RB1 models, each one more sophisticated than the last. "The RB1 will be an unbelievable personal robot," predicts RB Robot president Joe Bosworth. It will be a pet, but a pet that can turn on the television set, dial the phone—even be the phone.

In Sunnyvale, California, the creator of the Atari video games, Nolan Bushnell, is playing godfather to a couple of new personal robots produced by a company called Androbot. The first product, a little robot called Topo, can walk, talk, and deliver snacks in its optional Androwagon. Topo sells for less than $1000, but there's a catch: it is remote-controlled by means of radio signals, and the user needs a separate computer to give it instructions. These aren't really bad handicaps, since the radio controls work as far away as 200 feet—even through walls. And the robot can be used with any of the popular brands of home computers.

But Androbot is also producing a more sophisticated robot, B.O.B. Its name stands for Brains on Board. B.O.B. stands three feet high and rolls around at speeds of up to two feet per second (about 1.4 miles per hour) on a specially designed two-wheel system. B.O.B. finds its way around with the aid of five ultrasonic sensors. Two infrared sensors detect heat, so that B.O.B. can tell the difference between a person in its path and an inanimate object like a chair or table. (It can even distinguish between the heat of a human and that of a lamp or fireplace.) It has a voice synthesizer for talking. Like the RB5X, B.O.B. can sense when its battery is running low and can go to recharge itself.

Bushnell enjoys starting up new high-technology companies, and he loves to think up new applications. One of these is a robot travel machine. "Suppose I told you that you could have half the experience of a trip to Paris for one hundredth of the cost?" he says. He pictures getting into a little machine that is actually a robot, connected by a high-speed data link to another robot in Paris. "You can walk around in your robot and experience all the sights and sounds that the robot is experiencing on the streets of

Androbot's Topo robot is remote-controlled.

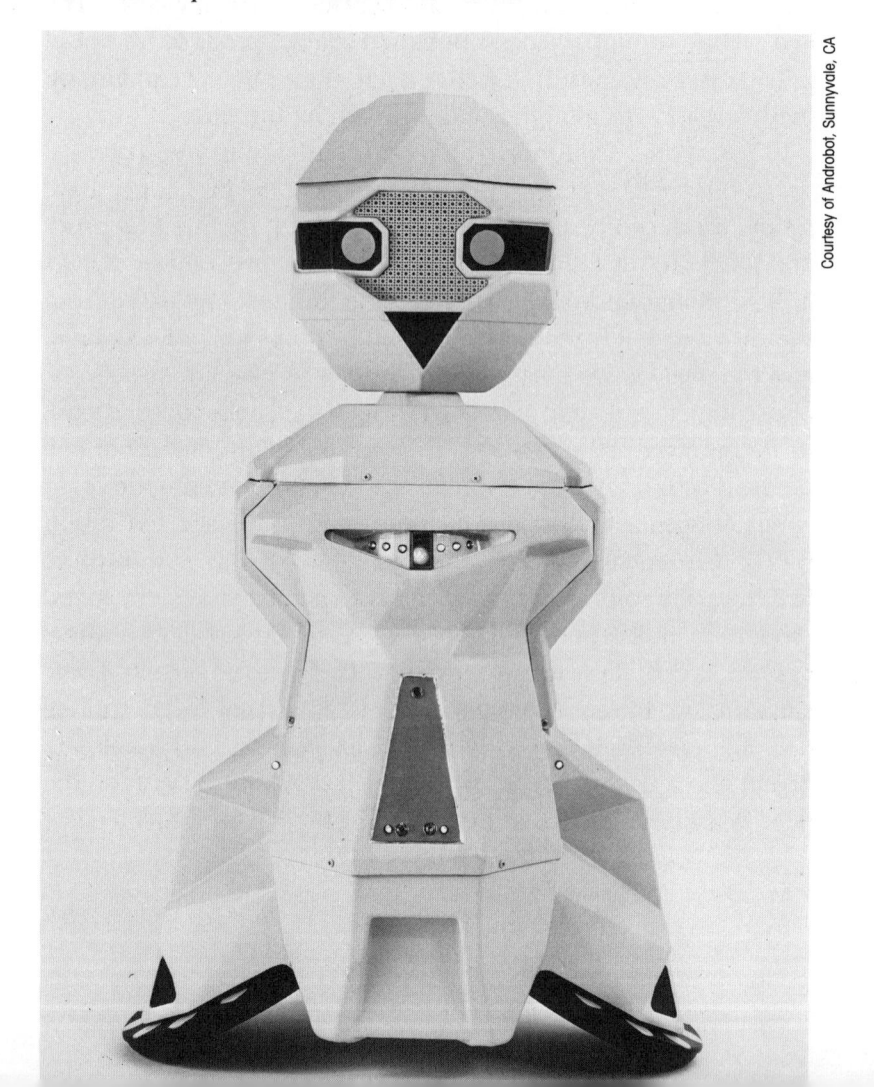

Androbot president Tom Frisina with his friend B.O.B.

Paris." It will be a long way from B.O.B. to a robot travel machine, but Bushnell thinks we'll have one someday.

Robots like Hero I, RB5X, and B.O.B. sound like fun to play with, but they are still far from the household robots of fiction. Do today's robot makers have the capacity to turn out a really useful home robot, or is that still in the future? "You'd have a tough time making Hero serve drinks, feed the dog, or wash the windows," says Douglas M. Bonham, the director of educational products at Heath. "There are a lot easier ways to wash the dishes than with this robot." Indeed, an article in the *Wall Street Journal*

85

But can it wash windows? A Unimate can.

about the new home robots was headed, "Hero the Robot Won't Wash the Windows, But It Will Happily Stroll Around and Chat." Yet robots can be built to do practical things like washing windows.

Back in 1980, the Merv Griffin Show featured a segment on robots. The star of the show was the Unimate manipulator. The robot arm was mounted in front of a closed window that was covered by a curtain. The robot reached around the curtain and pulled a cord to open it, then picked up a window cleaner and sponged water onto the lower pane. Then it turned the tool over to the rubber blade on the other side and "squeegeed" the window dry. The robot wasn't finished yet. It put the tool down, reached up to unlock the window latch, raised the window, and picked up a watering can to water the plants in the window box outside.

Like many industrial robots, Unimate is just an arm mounted in a fixed position. The confusing "real world" of the home needs a robot that can move around to get to the jobs that need doing without knocking over tables or stepping on the cat on the way.

The ComRo TOT can sweep the carpet and serve drinks.

Two new entries in the home robot race seem to be real household robots, not just pets, companions, or toys.

Jerome Hamlin of ComRo Inc. in New York has a line of household robots called the ComRo TOT. (ComRo stands for Computer Robot.) The three-foot-tall 50-pound ComRo TOT has a rotating head with a spotlight eye, two arms that can be controlled independently, and an ultrasonic ranging system. ComRo TOT can be remote-controlled by radio or can move around under the control of its own on-board microcomputer

87

"brain." It has a keypad for programming, or it can be programmed with cassettes from an Apple II computer. ComRo TOT comes with attachments such as a carpet sweeper, a tray, a lifting device (it can lift up to five pounds, compared to only one pound for Hero I), and a bucket. This is a robot that can really do household chores, such as sweeping the carpet and serving food and drinks. It can also tell jokes, greet guests, dance, and play obstacle games. It comes in various models, from a kit that sells for $1700 up to the deluxe model that costs $4595 and comes with a home entertainment center including a TV, radio, cassette player, and stereo speakers.

Hamlin started out building robots as props for TV (called show robots) and gradually grew interested in building "real" robots. He was actually the first to offer a household robot for sale. The ComRo I was featured in the 1981 Neiman-Marcus Christmas catalog as an ideal "His and Hers" gift. According to the catalog, ComRo I could open doors, serve guests, take out trash, bring in the paper, sweep, fetch, do light hauling, water the plants, dust, pick up after the children and pets, caddy at the putting green, and walk the dog. But its price tag was too high for the average family: the standard model cost $15,000, and the deluxe model, with color TV and stereo, sold for $17,500. Only three of the robots were sold, all to European buyers. Hamlin hasn't heard from them since and suspects they were robot engineers eager to take the robots apart and learn how they worked.

No household robots were offered in the 1982 Neiman-Marcus Christmas catalog, but that year's Hammacher Schlemmer catalog featured a different home robot on its cover. That was the Jenus, produced by Robotics International Corporation of Jackson, Michigan (which spells the robot's name Genus). The most sophisticated of the home robots then on the market, Genus also sported a high-class price tag of $7995. Production delays pushed the delivery date into 1983, but the dozen or so people who had ordered robot Christmas presents waited patiently.

The ComRo I household robot was a 1981 Christmas gift item.

Their lucky friends or relatives were due to receive a robot that could serve as a watchman, house cleaner, and butler, all in one. Built-in sensors could detect escaping gas, the heat or smoke of a fire, or the movements of a burglar. In an emergency like that, Genus would approach the danger, turn on a loud warning siren, and transmit a pre-recorded message to the local police or fire department.

Genus has an on-board computer that lets it learn its way around the house without a human to help it. Its ultrasonic ranging system keeps it from bumping into walls, furniture, people, or pets. Like the ComRo robot, Genus can be programmed to say literally anything. It can shake hands, play games, serve drinks, and vacuum the rugs. It comes with ready-made programs for many household tasks, but the user can also program in new abilities from a home computer or Genus's own built-in keyboard. In addition, it can be programmed to respond to spoken commands, and when its batteries are low it plugs itself into its own personal electrical outlet to recharge.

Computer experts and hobbyists are excited about the new personal robots, but for the average person who is used to the sophisticated robots in science fiction stories, they seem rather crude and limited. Probably most of the first home robots will be bought by hobbyists and people who like to be the first to buy anything new and expensive. Most of the people who buy robots just for status symbols will not have the knowledge and technical skills to operate their new toys properly, and those robots will soon be sitting in a corner gathering dust. But the hobbyists probably will quickly work out new applications for their robots.

Market researchers, who study people's buying habits, think that the first crop of personal robots are the first step toward really sophisticated humanlike robots. When those robots will arrive is still anybody's guess. Several advances will be needed before the average person will be as willing to go out and buy a robot as to get a new refrigerator or stereo set or car. First of all, home robots will have to be cheaper before they can appeal to large numbers of people. Secondly, they will have to be able to do more: how many people would be willing to pay a few thousand dollars for an animated toy or a walking burglar alarm—one that can't even walk up or down a flight of stairs? Perhaps most important, the robots will have to be more "user friendly"—able to follow simple spoken commands without needing complicated computer programming.

Most of the people who make industrial robots tend to be skeptical about the prospects for home robots. They think it will be a long time, perhaps fifteen years or so, before home robots are cheap enough and useful enough to become really popular. But it wasn't long ago that people were saying things like that about home computers, and robot hobbyists are much more optimistic. Gene Oldfield, owner of Robot Repair in Sacramento, California, thinks that the home robot could become as common as the personal computer within about two years.

Building Your Own Robot

In 1977 the computer magazine *Spectrum* sponsored an Amazing Micro-Mouse Contest, with a $1000 prize for the robot mouse that could find its way through a maze the fastest. More than 6000 people entered the contest and set out to design their own robot mice. Most of them found it harder to build a mouse than they expected, and only fifteen mice showed up for the race. Some of them got hopelessly lost and never managed to find their way out of the maze. The favorite, Catty-Wampus, had a very sophisticated computer brain but suffered from some design problems. It was so fast that it kept crashing into the walls of the maze and getting stuck. The winner, Midnight Flash, ran the maze in just thirty seconds, beating out the second-place Harvey Wallbanger by a full ten seconds.

What kind of people would enter such a contest? They were mainly students and engineers, ranging from teenagers to adults. What they all had in common was an interest in robots.

A generation ago, mechanically minded teenagers were likely to spend their spare time (and all the spare dollars they could

Courtesy of IEEE SPECTRUM. Photographs were taken in conjunction with the 1978 Amazing Micro-Mouse Contest, June 1978. © 1978 IEEE.

Here are some of the robot mice entered in the Amazing Micro-Mouse Contest.

scrape together) tinkering with cars or building their own radios from spare parts. Today's kids are keen on electronics, and robots have the biggest appeal. Building a robot can also be fascinating for an adult hobbyist: there is the fun and interest of developing the robot's brain, motors, and sense systems, and the thrill of creating something that is a little like life. The robot builder may also be inspired by the hope of big profits: if the robot is good enough, maybe it can be marketed and sold by the millions! At the least, the proud builder may be able to show off the robot at parties or supermarket openings. Perhaps a newspaper or magazine will pick up the story and feature a picture of the robot and its creator.

There have already been a number of success stories in the robot hobby field. For instance, Tod Loofbourrow's interest in robots was sparked by a visit to the Expo '67 World's Fair in Montreal, Canada. Tod was only six then, but he was fascinated by a display on robots and wanted to build one. He started building a robot about nine years later, with ideas from electronics magazines. He started with a simple frame, added a car battery and a power system, and then continued one step at a time, adding one system after another. Eventually Tod's robot, which he named Mike (short for "microtron"), was equipped with a multiple-speed moving system and an ultrasonic ranging system (like bat sonar) to help it find its way around. Bouncing sound waves showed it the position of objects in the room, but it also had eight bumpers to signal when it bumped into things. Tod built a voice recognition system into his robot and taught it to obey simple commands. "I can call its name and it comes," he says; "tell it to go left, it goes left; right, it goes right."

Building the robot wasn't the end of the story for Tod Loofbourrow. He gave a talk about it at the Amateur Group of New Jersey. After the talk he was amazed to receive an offer to write a book about his project. He wrote the book, called *How To Build a Computer Controlled Robot*, when he was sixteen, and before long he was a published author.

Jerome Hamlin, an independent filmmaker and TV producer, progressed from building model airplanes to building robots. He made his first show robot, Bumpy, when he was hoping to get a job building robots for the movie *The Empire Strikes Back*. The producers of the movie didn't buy Bumpy, but Hamlin was hooked on robots. He built others, such as Bubble Bot, a remote-controlled puppet that moves its lips, hands, and arms and wags its head. Soon Hamlin was in business building robot props for movies and TV. But he had ideas for some "real" self-controlled robots, too. Along with ComRo I, the first general purpose household robot, he produced a pet robot "dog" named Wires. Hamlin hopes that his newest household robot, the ComRo TOT, will be a big seller, but Wires has been discontinued. "I sold several through the Nieman-Marcus catalog," he says, "but they weren't cost effective. They went for $650 each, but they cost me

Here is a partly finished TOT being built at the ComRo workshop.

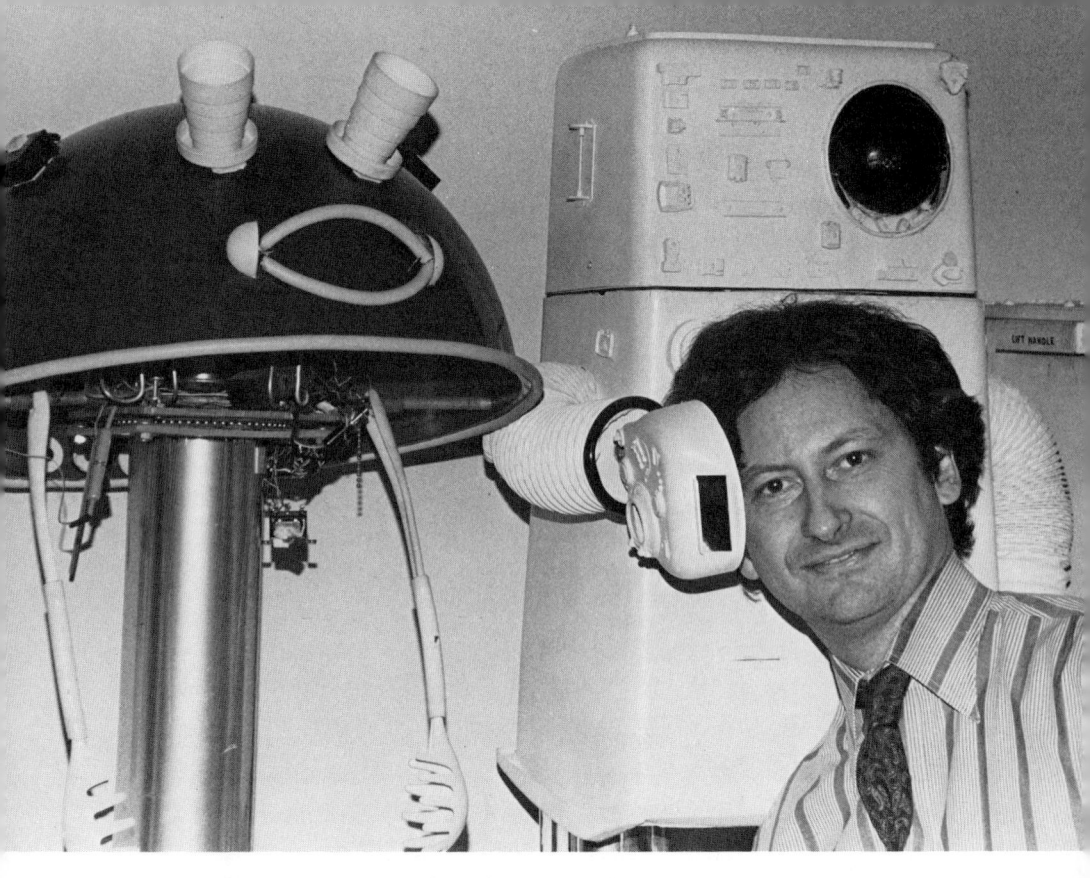

Robot builder Jerome Hamlin, with Bubble Bot and Bumpy.   Robert Alan

more than $400 to build—not including labor! Of course, they'd be cheaper if they were mass produced."

Frank DaCosta, a computer technician and robotics hobbyist, found another way to make a profit from a homemade robot dog. Like Tod Loofbourrow, he wrote a book about his project, *How to Build Your Own Working Robot Pet*. The project started with a half-joking conversation. DaCosta and his wife were discussing the fact that their apartment complex would not allow them to own a pet dog. "Well," he said, "if we cannot have a *real* dog, we'll just build one—a robot dog!" The robot he built didn't look much like a dog, but it could roam around the apartment (on wheels equipped with a Soniscan system to keep it from bumping into things), bark (with a voice synthesizer), and obey simple spoken commands. The DaCostas saved a lot on dog

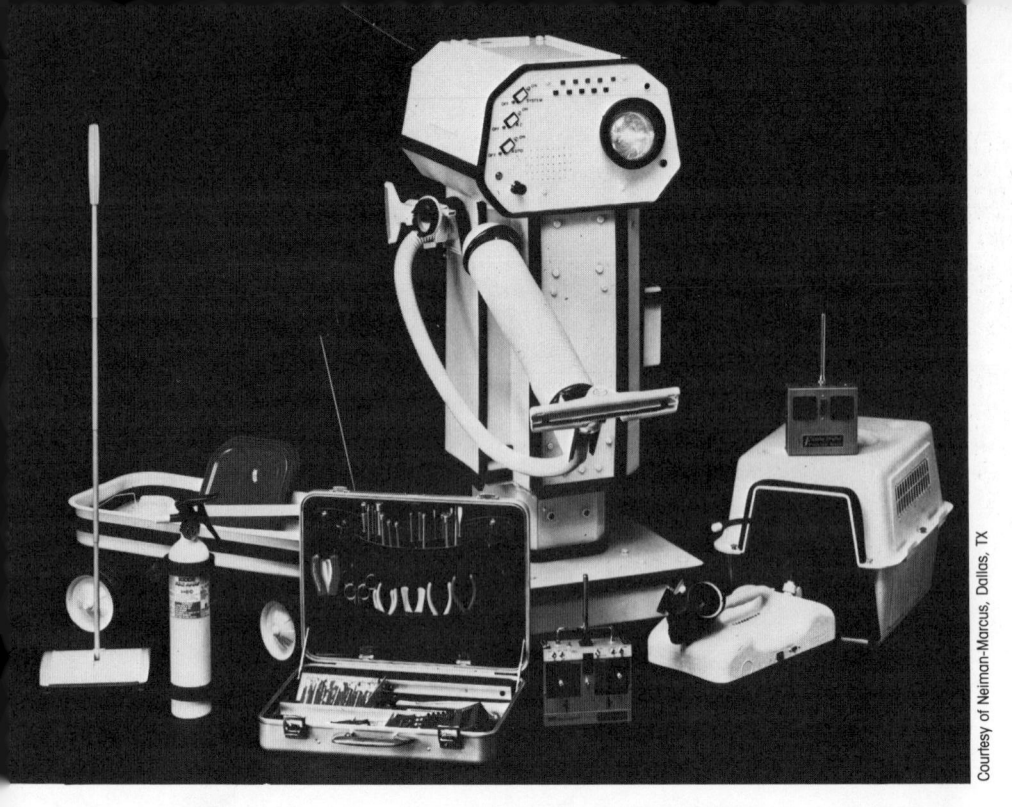

The Neiman-Marcus catalog showed the ComRo I with all its attachments for household chores. At right is the robot "dog" Wires, with its carrying case.

food, but their pet did need regular "feedings" of an electric charge for its batteries.

Physicist Gene Oldfield built his first robot, Entropy, while he was in graduate school. Entropy had wheels, a computer brain, and a sonar system that helped it to navigate through doorways without bumping. Oldfield moved on to an electronics job in Silicon Valley (an area near San Francisco where many high-technology companies are located) and then a teaching position in a university, but meanwhile he kept on building robots. His latest project is a new personal robot that aims to be "a programmer's delight." Called the ROBOCYCLE or "Bike Bot," Oldfield's new robot is built for speed. It can go six to nine miles per hour (compared with only 1.4 mph for Androbot's B.O.B.), find its way around with a sonar system that gives distance and

Gene Oldfield in his robot repair shop. His first robot, Entropy, is in the foreground. Chad Chadwick © 1983

angle, and learn the layout of its environment. ROBOCYCLE has sensors for motion detection and a 10-word voice input. Oldfield plans to produce only twenty ROBOCYCLES, for sale (at about $2000 each) mainly to computer specialists who can add new programming ideas. A bimonthly newsletter will help the robot owners to share the new programs.

Robots have become a full-time occupation for Gene Oldfield now. He has launched a new Robot Repair business in Sacramento, California. With home robots coming onto the market, there will be plenty of need for his new services.

John Blankenship, a computer technology professor in Atlanta, Georgia, whose robot TIMEL was featured in a series of articles in the magazine *Robotics Age*, tried to use inexpensive everyday items in building his robot's body. That way he could save his main investments for the computer brain and the various systems connected to it. TIMEL, whose name was formed from the initials of Truly Intelligent Mechanical Electrical Life, had a plywood base, a cardboard body shell, and a fishbowl head spray-painted with metallic paint. (It glowed nicely when lit from inside.) Its forearm was made from a section of tube from the center of a roll of carpet, with a potato-chip can to form the wrist. "The cardboard-to-cardboard bearing—with a little Vaseline added for smoothness—may not pass an industrial inspec-

TIMEL's body was built from inexpensive materials.

Contest-winning robot Avatar is a three-dimensional sculpture with brains.

tion," writes Blankenship, "but it works efficiently and never seems to wear down."

Blankenship found that TIMEL wasn't quite intelligent enough to be really challenging. He has plans for building a much improved TIMEL with an Apple computer on board. Meanwhile, he has been working on ways to turn a whole house into a "robot" that accepts voice commands and talks back to its owners in its own synthesized voice. The system, controlled by an Apple computer, controls the heat, telephone, security alarms, and even wakes Blankenship up in the morning. He has written a book about this project, called *The Apple House*.

The winner of the 1981 *Robotics Age* "Homebuilt Robot Photo Contest" was a robot named Avatar. Its creator, Charles Balmer, Jr., recommends that robot hobbyists should emphasize a general purpose design, with plenty of flexibility. "Don't attempt a design that is too ambitious," he advises. "If your robot lab is designed properly, you will be able to expand its capabilities as

99

time goes on." He calls a robot a three-dimensional sculpture. Building one provides scope for the imagination and artistic sense but also needs a good balance of practicality. "A robot is somewhat like a child," he says. "It requires a great deal of patience, time and energy to construct, and then as it limps and crashes and smokes its way to adulthood, we as mothers and fathers learn something about being a robot while hopefully our robot learns a little bit about being a human." Balmer spent five years building his robot, and is currently developing software—the programs that make the robot work. Now he is thinking of producing a line of personal robots for sale.

Another notable homebuilt robot is ROBART, a five-foot four-inch creation of Bart Everett. ROBART is an electronic watchman who patrols the Everett home, uses its sound, light, heat, and motion sensors to detect intruders, and shouts a challenge to them with its electronic voice. Everett built ROBART in 1980 while attending the Naval Postgraduate School. The publicity the robot received led to Everett's assignment as Assistant for Robotics for the Naval Sea Systems Command in Washington, D.C. The original ROBART and a more sophisticated version, ROBART II, are now doing double duty. When they are not on home patrol, they travel to the Naval Surface Weapons Center in White Oak, Maryland, where they serve as design aids in a project for developing shipboard applications of robotics.

Robart is all machine, with its interesting electronic innards clearly visible through plexiglass panels. Canadian hobbyist Donald Dixon has a different idea of what a robot should look like. He created a robot called Ahmad, with a latex-rubber face molded to resemble his own features. The computer-controlled Ahmad rolls around the Dixon house on three wheels, scolding Dixon when he is tempted to take a midnight snack and giving messages to his wife. At night Ahmad is stationed outside the children's room. "Go back," the robot tells them if they try to go out, and if they don't obey, Ahmad wakes his master with an alarm.

After building the electronic sentry ROBART (left), Bart Everett built an improved version, ROBART II (right).   H.R. (Bart) Everett

How much does it cost to build a home robot? It depends on how ambitious the project is. Tod Loufbourrow spent about $450 to build Mike, but the prices of electronics equipment have been dropping since then. He estimates that he could probably build a robot for about $300 now. The prize-winning Avatar, on the other hand, took an investment of about $2000, but its creator, Charles Balmer, Jr., estimates that it might have run as much as $100,000 if he had not been able to use surplus parts!

Industrial roboticist Joseph Engelberger has this to say about robot hobbyists: "Oh, they can have a whole lot of fun. I think everyone has a chance to make a contribution."

Robots for Fun and Games

In 1977 an exciting announcement came from a small company in Rutherford, New Jersey. Anthony Reichelt, founder and president of Quasar Industries, said that his company would begin production of "Domestic Androids" late in 1979. These would be real household robots, able to scrub and vacuum floors, answer the door and the telephone, serve snacks, and entertain the children with rides, dances, taped stories, and games of hide-and-seek. All this for only $4000.

The announcement stirred up a storm. Computer experts claimed that real home robots could not be produced for many years, and certainly not to sell for $4000. Reichelt's robots, which were being used as novelty demonstrations at trade fairs and exhibitions, were nothing but hoaxes, the experts said. They weren't really robots, but merely remote-controlled toys. Perhaps that was true of the show robots, Reichelt admitted, but Klatu the Domestic Android would be different. The show robots that his company had been producing since 1969 were merely a way to make money to finance the production of real working robots. Powered by an energy cell and steered by a sonarlike system feeding into a small computer in its midsection, the robot Klatu would be the real thing at last. It would have a vocabulary of about 250 words and recognize 50 spoken commands. An early problem in the word-recognition system led to Klatu's name. When its inventor, Reichelt, activated it and commanded, "You talk," the robot repeated what he said—backward—as "Klatu."

It made a good story, and even the controversy added to the publicity. But 1979 came and went; Quasar announced a few postponements and finally stopped promising to produce Domestic Androids. Reichelt was ahead of his time, though not nearly as far ahead as the experts thought. Working household robots were already for sale two years later, and just a year after that they were down to the price range of the overdue Klatu.

Quasar's "Domestic Android" Klatu was a bit ahead of its time.
Quasar Industries, Inc.

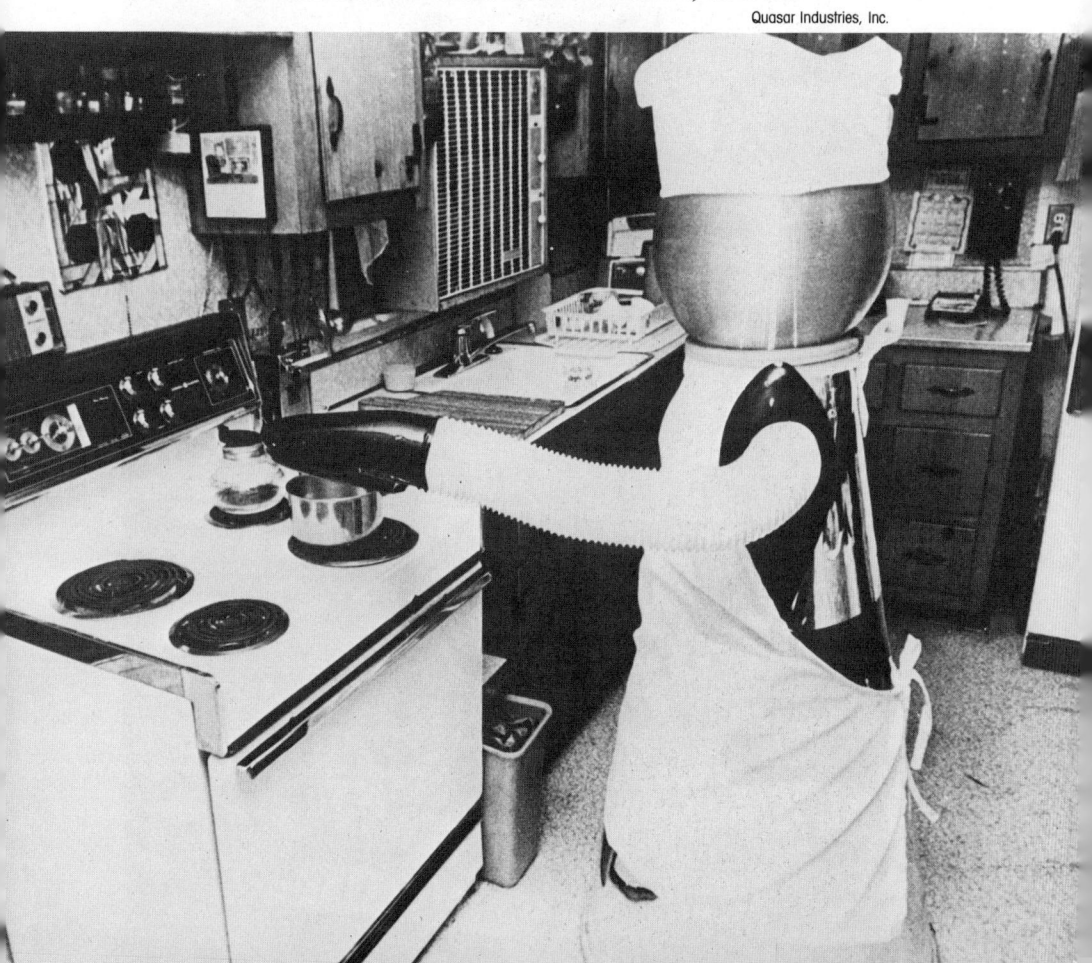

Quasar Industries didn't come through with the Domestic Android, but they continue to produce a line of highly successful show robots. These mechanical creatures enliven openings of banks and supermarkets, meetings, fairs, and trade shows.

Quasar is only one of the companies active in the thriving business of entertainment robots. The Robot Factory of Cascade, Colorado, produces a line of robots including a squat little cylinder called Six Robot; a big, squared-off model called Ralph

Remote-controlled Andrea Android was designed with a very human appearance.

Android Amusement Corp.

Ralph Roger Robot greets people by remote control.

Roger Robot that walks, talks, shakes hands, and tells jokes; and the Humanoid Robot, which looks like a science fiction creature from outer space. These robots are all remote-controlled, but the furry Hot Tots can also work automatically to put on a pre-programmed show that includes riding a bicycle and playing the banjo, guitar, and keyboard.

Andrea Android and Adam Android are two attractive human-like "dolls" designed by Gene Beley of Android Amusement Corporation. They move and speak by remote control. Android Amusement Corporation also produces a Drink Caddy robot, models of which have appeared in *Playboy* Magazine, starred on an episode of the CHiPs TV series, and even have been arrested in Beverly Hills. (The delinquent robot, remote-controlled by Beley's teenage sons Shawn and Scott, was turned off and hauled off to jail after it passed out business cards on a Beverly Hills street and refused to identify itself to investigating police officers. The boys eventually bailed it out by paying a $40 towing fee.)

Gene Beley believes that the first home robots will be more for entertainment than for practical use. "It's nice to say you're going to make a home robot that's going to do all kinds of wonderful things," he comments, "but if you ask someone if they'll spend $20,000 for it and they say: 'Are you crazy? I can buy a vacuum cleaner for $200,' it doesn't make sense." For quite some time, he believes, home robots will be toys for wealthy people. "I think we're selling fantasy. We're talking about bringing a miniature Disneyland into your home." Eventually, though, robots will have more practical uses. "When the Wright brothers first flew, everybody said: 'That's cute. That's nifty. But what are you going to do with it?' Uses will be found for robots the same way they were found for airplanes."

Nolan Bushnell, video game inventor and sponsor of the new Androbot robots, has another robot enterprise going. His chain of Pizza Time restaurants features a combination of fast food, video games, and entertainment provided by giant singing robot animals. A customer buying pizza and drinks receives a number of tokens that can be used to play video games or set the robot entertainers singing. The performing robots include a quartet of dogs called The Beagles that strum guitars and sing old Beatles' hits.

Big Trak is a toy tank with a tiny computer brain.

Some robot experts criticize the show robots as a kind of make-believe that gives people unrealistic ideas about what robots really can do. But they are only meant for fun.

Another form of robot fun is provided by the many robot toys. They range from windup toys in robot shapes to remote-controlled devices that are very close to real working robots.

Milton Bradley's Big Trak is a programmable toy tank with a microprocessor control. (A microprocessor is a tiny computer on a silicon chip.) The user taps out commands on a keyboard and can enter a sequence of up to 16 commands. These instructions tell the tank how far and how fast to go and when and where to turn. When the toy came out, computer experts were amazed that a toy with a tiny computer brain could be produced and sold for only $40. The key is mass production, geared to selling a million of the toys a year. An article in *Robotics Age* reported that Milton Bradley spends more than $600,000 each year for research on robot and computer toys—a sum that is about equal to the entire yearly budget for robotics research at NASA!

Another robot toy that is very close to the real thing is the Armatron, a manipulator arm imported from Japan by Tomy Toys of California. Armatron is remote-controlled with joysticks that work by means of gears. With these controls, the user can rotate the whole arm, raise and lower it, bend it at the elbow, raise and lower the hand, rotate the wrist, and open and close the hand. These are capabilities found in experimental robotic arms costing about ten times as much as the toy, which sells for less than $50. An article in a recent issue of *Robotics Age* discusses ways to adapt the Armatron to computer control, converting it to a real robot.

It's hard to tell how to classify the new robot kits distributed by Robot Shack of El Toro, California. The seven-inch DROID-BUG comes in the form of a ready-to-put-together kit that costs less than $100. Powered by four C batteries, it scoots around the floor, senses obstacles with a feeler, then turns away and continues in a new direction until it meets another obstacle. DROID-BUG makes a loud buzz like a hurt bug whenever it senses something in its path.

DROID-BUG's bigger brother X-1 is a more sophisticated robot that moves around controlled by an on-board control panel. Optional parts that can be added to the basic kit include a hearing sense that turns robot functions on and off in response

The Armatron is a toy manipulator very similar to real robot arms, but it is controlled by joysticks.

ROBIE the personal robot (left) has male and female faces. X—1 and DROID—BUG (right) can be built from kits.
Robot Shack, El Toro, CA

to voice commands; ultrasonic and heat sensors that detect the approach of humans or animals and equip X-1 for sentry duty; a chirp horn alarm to frighten intruders away; and a water gun that squirts water out of the robot. (The water gun and chirp horn alarm can both be connected to the detector system.)

Robot Shack was founded by aerospace scientist Eugene F. Lally, who wanted to provide hobbyists with the means for building personal robots for both fun and learning. His main goals were to make robots both easy to build and cheap enough to be affordable. His most ambitious project is the PR-1, or ROBIE, a personal robot with some intriguing gimmicks. It has a male face on the front and a female face on the back. ROBIE can "hear" and respond to spoken words, either by "talking back" in a synthesized robot voice, or by turning the robot functions on or off. Its scanning antenna sticks up from its head and spins, and its eyes, nose, and heart flash on and off. ROBIE can blow up a balloon through its belly button and shoot a water pistol.

Another option is a heat-sensitive infrared target and a gun to shoot at it. If you hit the robot on the bull's-eye, it says "Ouch!" Whether the Robot Shack robots are toys or personal robots, they certainly sound like fun.

Also fun and definitely a robot is the Chess Robot Adversary that is sold through the Hammacher Schlemmer catalog for $1500. Unlike the mechanical Turk that had a human chess player hidden inside, this robot really can play chess. It can be programmed to play against a human opponent or to demonstrate classic chess games of the past. It is built in the form of a chess board with a mechanical arm that moves the chess pieces. (The first ones shipped had some problems with the arm, which sometimes knocked the pieces over or moved them to the wrong squares, but these problems have been solved with some fairly simple adjustments.) The robot doesn't look human, but it has been programmed to make some very human-sounding cheers and groans when the game is going well or badly.

As computer chips, voice synthesizers, and other electronic systems continue to get cheaper and better, more and more toys and games probably will go robotic.

The furry Hot Tots can put on a preprogrammed show.

Robots vs. Humans: Where Do We Go from Here?

We are just at the beginning of the robot age. "Intelligent machines" are just starting to affect our lives. In the years to come, more capable and versatile robots will be developed and will penetrate into more and more areas of daily life.

Household robots available today will probably appeal mainly to hobbyists, but it won't be long before increasingly useful models will be developed. For awhile they will have hot competition from "smart machines" that do just a single household job, such as an automatic vacuum cleaner that cleans the house without human supervision. But gradually the idea of having a single appliance that can do many chores and be a pet and companion besides will appeal to more and more people—especially since a robot could plug itself into a TV set and double as a home computer. Eventually home robots will probably be as widespread as televisions, telephones, and automobiles.

Robot vision systems now being developed will eventually be sensitive enough and fast enough to use in automatic systems for driving cars. With automotive engineers already beginning to

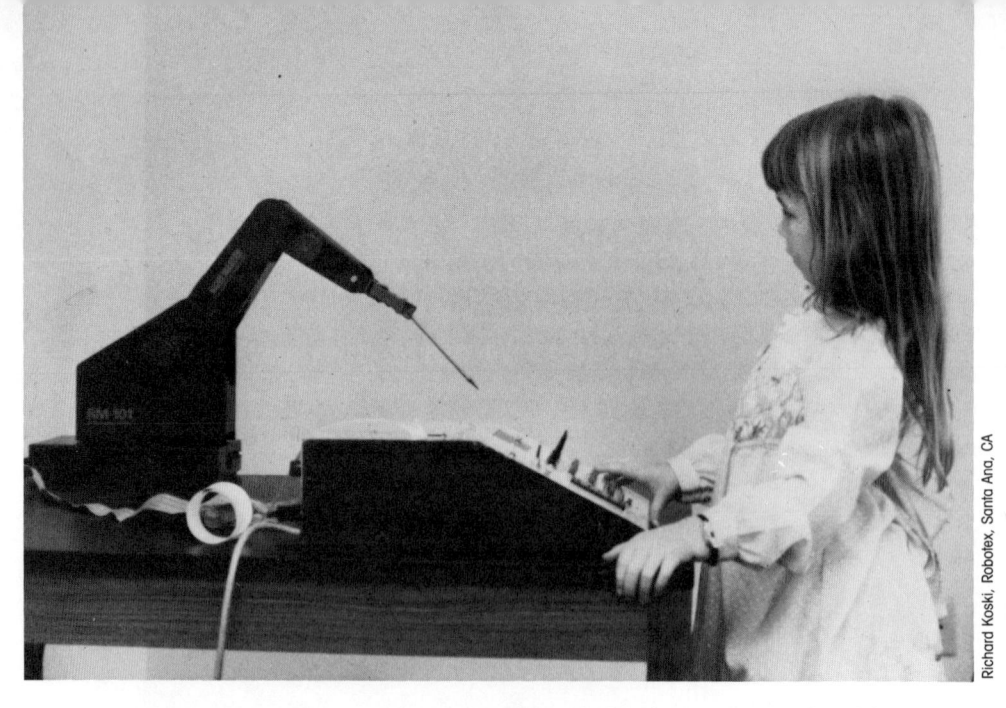

Children who grow up with robots will probably feel more comfortable with them than many adults do today.

computerize many of the internal workings of cars, it seems likely that the family car of the future will be a mobile robot able to pull out of your driveway and take you to the destination you punch into its computer keyboard, automatically selecting the best route from its memorized road maps, avoiding collisions with cars and pedestrians along the way, and even pulling into an automated service station when it senses that its fuel (or perhaps its electric charge) is running low.

Partly and fully automated factories will be much more common, of course, but robots will also be penetrating into other job areas. There will be robots waiting on tables in restaurants, robots working as lumberjacks, and robot farmhands automatically spraying, cultivating, and harvesting crops. In offices, robot secretaries with voice recognition systems will faithfully transcribe and type correspondence, which robot messengers will deliver to the nearest automated mail processor. Some researchers who specialize in the study of future trends think that there will even be robot managers to make decisions about the policies of the company. A La Jolla, California, firm called General Robotics Corporation has already developed a computer program to act as an "electronic jury," evaluating court cases on the basis of the evidence and legal precedents.

What will *people* be doing if robots will be doing all these jobs for us? That is a hard question, and the answers are very uncertain. Some people fear that we will handle the problem foolishly, letting robots displace human workers without finding new places for these people. They think that for many people the spread of robots will result in unemployment, poverty, and misery. Others have confidence that we will be wise enough to use robots to make our life easier, to reduce drudgery while expanding our leisure time and devising new and interesting things to do with this extra time.

As robots become more common, how will people react to them? In many factories, human workers have become very fond of the robot coworkers, but some people have been angry and suspicious. They have seen so many stories about robot menaces on TV that they think these mechanical creatures must be untrustworthy.

No one has yet programmed Asimov's Three Laws into a working robot—we don't know yet whether that will ever be possible. But it seems certain that no robot has ever yet hurt anybody on purpose. That does not mean that robots aren't dangerous. In fact, they may be most dangerous when people get used to them and learn to accept them. When you work next to a robot on an assembly line and see it performing faithfully day in and day out, carrying out exactly the same actions in a perfectly repeatable pattern over and over again, you come to depend on its continuing to do so.

But robots are machines, and machines can malfunction. In one reported case, a robot researcher watched in horrified amazement while a robot suddenly began to rip itself apart. Even when a robot is working perfectly, it can cause harm if people nearby aren't careful. An industrial robot is big and powerful. If you walk into its work space, it may not realize you are there and may hit you accidentally. There has already been a case, in Japan in 1981, in which a robot killed a person. It wasn't the robot's fault. The human worker climbed over a guard fence, attemped to make an adjustment on the robot without turning off its power, and was crushed by the robot's arm.

How people react to robots may be affected by what the robots look like. Perhaps when designs become more sophisticated and robots can be made to look more human, people will accept them more easily. Or perhaps the reverse may be the case. Many people become quite fond of the machines they work with, even though they don't look at all human, and most people think an R2-D2-type robot is "cute." Perhaps the robots that still look like machines will be more easily accepted in the future, while robots that look too much like humans will be perceived as threats. People may regard them as competitors.

It seems likely that we will be able to make future robots look human (if we want to), but will they ever become as smart as humans? Will robots ever be able to really "think" and come up with creative new ideas? Industrial robot pioneer Joseph Engelberger doesn't think so. "Robots never will have the broad general intelligence and sensitivity of humans," he says. "We'll never have the ability or the economic incentive to make robots that do abstract painting or lead their team to victory in the Super Bowl."

Some artificial intelligence researchers might disagree. They believe that as computing power increases and computer "brains" grow as complex as a human brain, something like human intelligence will develop in the machines. They are trying to develop programs that mimic various aspects of intelligence, and they have made some impressive advances.

Take the robot chess player, for example. Playing a game like chess successfully requires thinking ahead, to figure out what the other player may do and what choices will be left in each possible case. With their enormous memories and high speed, computers have some natural advantages in that kind of activity. Even so, philosopher Hubert Dreyfus wrote in 1964 that computers would never be able to equal human intelligence. He commented scornfully, "No chess program can play even amateur chess," pointing out that a ten-year-old had beaten the best of the current computer chess programs. Less than two years later, an MIT graduate student, Richard Greenblatt, devised a chess-playing computer program called MacHack and challenged Dreyfus to a chess match—and the computer won.

Is playing a rather mathematical game like chess according to a program designed by humans really thinking? It is hard to say. But another computer was designed to learn the rules of one game and then to apply its experience to learning other games more easily; that computer was certainly doing something resembling human mental activity.

Some learning machines have been taught to recognize patterns, like letters of the alphabet or pictures of people's faces. The UCLM II, in London, was taught to recognize faces when it was presented with only a fragment of the pattern, such as the top of a person's hairdo or just the mouth and chin. ACRONYM, a similar computer program at Stanford University in California, can identify jet airplanes after a single fragmentary glimpse. But so far no one has been able to teach a pattern-recognizing program to generalize well enough, for example, to tell whether a particular face is that of a man or woman.

Computers have been programmed to draw pictures, compose music, and even to write poetry. One computer, carefully primed with a good-sized vocabulary, the appropriate rules of grammar, and rules for constructing poems came up with the intriguing couplet: "O panic not to this docile juice/Finally few of my jackets did distrust the goose." Another computer program, PARRY, is designed to sound like a mental patient. Conversing with a human, the computer seems to respond to the human's comments, but its replies often have an emotional overtone. It might respond to a factual statement with a comment like, "YOU

ARE MAKING ME NERVOUS," or remark abruptly, "THEY ARE OUT TO GET ME." If PARRY's answers don't always make sense, then they are distorted in a manner very similar to the disordered thinking of someone who is mentally ill.

A more rational program, designed by Patrick Wilson at MIT, deduced the characteristics of an arch by studying various types of bridges. The same program read Shakespeare's plays *Hamlet* and *Macbeth* and concluded that "a weak man married to a greedy woman is likely to be evil."

Programs like these are very intriguing. Are they merely clever tricks, or does the computer understand what it is saying?

So far there is probably very little (if any) real understanding. There are many things we do not yet know about how we humans use and understand language. Stanford Research Institute artificial intelligence researcher Barbara Grosz cites the classic "Monkey and Bananas" problem to illustrate some of the problems involved.

A young monkey is standing under a banana tree. A bunch of bananas dangles above its head, too high to reach. Then an older monkey comes by. "I'm hungry," says the young monkey. "There's a stick under the old rubber tree," the older monkey replies. The older monkey understands not only the words the young one has said, but also the problem it is trying to convey. The older monkey's answer assumes that the younger one will understand that the stick can be used to knock down some bananas.

If the young monkey said the same thing—"I'm hungry"—to one of today's robots, the robot might answer, "I understand." It has recognized the words and related them to their meanings in its memory bank, but it has no understanding of the deeper meanings—that the young monkey is asking for help.

The young monkey is frustrated by the robot's answer. "Can you help me get a banana?" it adds. "Yes," answers the robot. It has now recognized the young monkey's problem, but it has

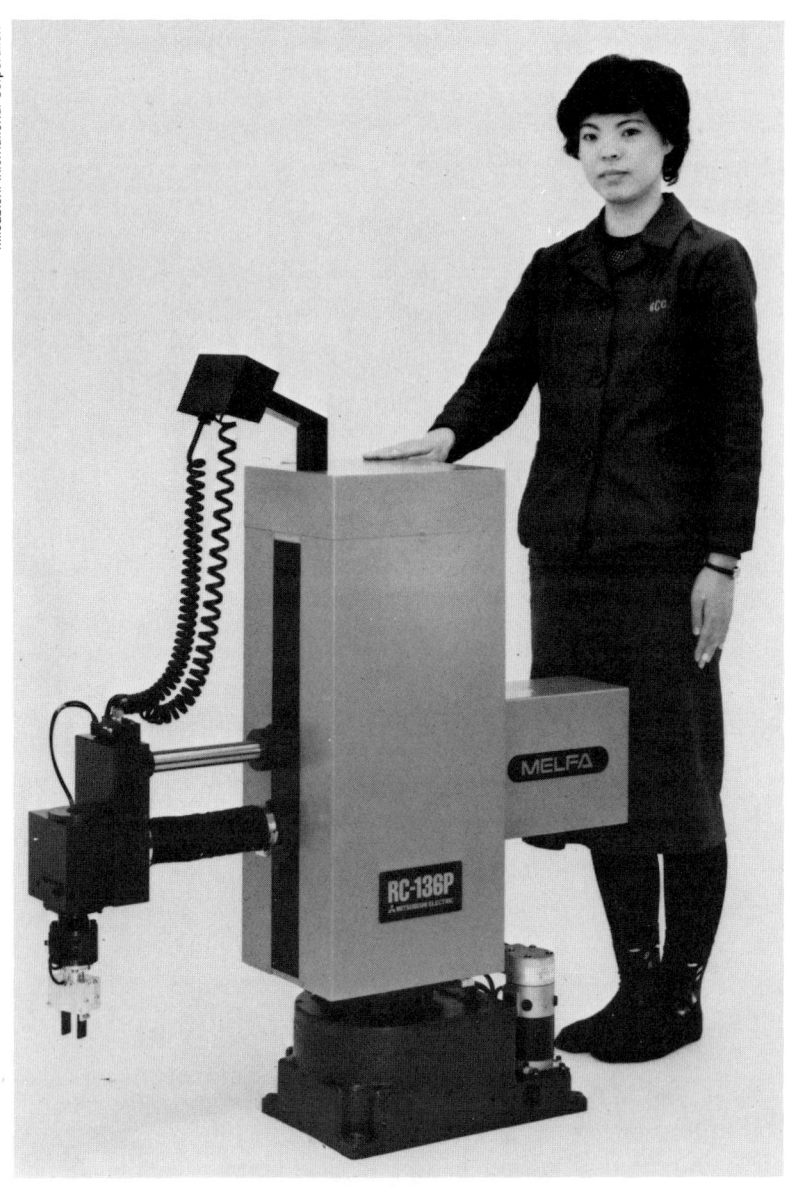

Today the Japanese are building rather limited industrial robots. But their Fifth Generation Computer Project will produce robots that are much more like humans.

taken the question so literally that its answer is not very helpful.

Artificial intelligence researchers are trying to make their computer programs more intelligent, so that if the young monkey said, "I'm hungry," the robot would answer, "There's a stick under the old rubber tree. You can use it to knock down a banana." Whether they will reach their goal and make robot brains that can really "think," only time will tell.

Japanese researchers are betting that they will. In 1980 the Japanese Ministry of International Trade and Industry announced the launching of a giant effort, the Fifth Generation Computer Project. Computer science has made such enormous strides in just a few decades that people in the field tend to talk about computer development in terms of "generations." The first generation consisted of the original 1940's computers built with vacuum tubes. The second generation was ushered in by the transistor and the smaller, faster computers it made possible. The silicon chip yielded further advances in the third generation. Large-scale integrated circuits brought today's fourth generation. Tiny but complicated microprocessors that are literally a "computer on a chip" are being built into a variety of "smart machines" from smart ovens and automobile engine controls to sophisticated scientific instruments that not only measure but even run whole experiments by themselves, without the need for human supervision. The fifth generation computer, as seen by the Japanese, will pack a million transistors on a silicon chip the side of a pinhead. It will be equipped with sensors to permit it to hear and see, and it will be able to talk to people, ask and answer questions, learn, and put facts and ideas together to form new ideas—very much the way people do, in fact.

According to Michael Dertouzos, director of the Laboratory for Computer Science at the Massachusetts Institute of Technology (MIT), this kind of computer would have fantastic applications. You could carry around a little computer "bug" with eyes and ears that would observe (and *remember*) everything you read, everyone you met, and exactly what they said. Months later, you could ask your bug about a particular conversation and have it immediately recalled to you in perfect detail. A factory manager could tell a computer, "Increase production by 10 percent," and the computer would figure out how to do the job—and do it. In Japan the effort to build fifth generation computers is a national research effort, with both government and private industry cooperating. Artificial intelligence researchers in the United States are not as closely coordinated, but they too are working toward similar goals.

Meanwhile, futurists continue to speculate on the future of robots and their relationships with humans. Arthur Harkins, director of the graduate program in futures research at the University of Minnesota, suggests that some day people will "marry" robots that look like humans and are programmed to speak and act in humanlike (although rather unimaginative) ways. The people who make such marriages will probably be those who cannot easily find human companions—for example, old people, those with severe mental or physical handicaps, or people in prison.

Ira Levin's science fiction horror story *The Stepford Wives* told about a group of men who murdered their wives and replaced them with cleverly made robot duplicates. The men of Stepford thought their new "wives" were ideal: they were perfect housekeepers and attentive wives and mothers, completely devoted to their husbands and children. They never argued with their husbands and were incapable of doing anything selfish, thoughtless, or unpredictable. Few real people would ever be satisfied with a wife (or husband) so thoroughly dull and "robotlike." But as computer sophistication increases and robot brains grow more and more like human brains, robots may develop complex personalities. They may become capable not only of imagination and originality, but even of emotional reactions. When robots can be made not only to look like humans but also to act and react like us, the distinctions between robot (or android) and human will begin to blur. We can imagine a future in which robots organize in activist groups to lobby for "robot rights," while some try to "pass" as humans.

Another long-range speculation is that current research on "biochips," semiliving molecular-sized circuitry, will ultimately transform the human race. The first step will be implanting biochips to fix faulty organs and restore lost functions like vision in the blind or movement in the paralyzed. But the biochips will also provide a way of increasing our brainpower enormously. Eventually, people with biochip implants may become robotlike cyborgs—part human, part machine. We may even achieve a kind of immortality: if a person has a biochip electronic brain, the contents of the brain—the person's personality—can be transmitted to a receiver and preserved, even if the original body and brain die.

The robot age should be an exciting time to live!

Glossary

android an artificial creature with a basically humanoid form, especially one made from biological materials.
artificial intelligence (AI) the field of research that tries to duplicate human thought processes by computer programs.
automation the operation of machines or systems by mechanical or electronic devices that take the place of humans in making observations, movements, and decisions.
automaton a mechanical device capable of moving on its own.
CAD/CAM Computer Aided Design/Computer Aided Manufacturing: a system in which computers are used to design objects and then control the operation of automated machine tools that build them.
cathode ray tube a vacuum tube in which high-speed electrons produce a glowing spot on a fluorescent screen.
chip a small piece of specially treated silicon that forms an integrated circuit.
circuit board a board on which various electronic elements are assembled.
computer a programmable electronic device that can store, retrieve, and process information.

electrode an electric conductor (usually made of metal) through which electric current enters or leaves a medium (such as a solution, a gas, or tissues of the body).

feedback the return of part of the output of a machine or process to its input, where it may influence the further operation of the system.

integrated circuit a tiny complex of electronic parts and their connections, produced on a small slice of material such as silicon (a chip).

manipulator a mechanical device for handling and moving things.

microprocessor a central processing unit of a computer that is produced on a single integrated-circuit chip (or on several chips).

phoneme a sound element of speech.

photoelectric cell a device in which light is transformed into electrical signals.

pixel picture element: a small segment of the field of vision in a robot's vision system. Each pixel is rated according to its brightness or darkness. Together the pixels form a pattern of dots that yield a picture whose sharpness of detail depends on the number of pixels and shades of difference the system can distinguish.

program the set of working instructions for a computer or robot.

prosthetic device an artificial body part or organ used to replace one that has been lost or paralyzed.

remote control control operating at a distance. Manipulators can be remote controlled through mechanical or electrical linkages or by radio signals.

robot a manipulator that works according to programmed instructions to perform a variety of tasks; it can be reprogrammed for different tasks.

robotics study of the construction, maintenance, and behavior of robots.

sensors devices providing robots with "senses" for perceiving their environment, such as vision, hearing, and touch.

servo control an automatic feedback system for controlling mechanical motion by continual small adjustments, based on information on the position and speed of the parts involved.

show robots robots designed for entertainment. Often these are remote controlled, and they may not be reprogrammable.

silicon an element that occurs in sand and rocks. Very pure silicon with tiny amounts of carefully controlled added impurities is a semiconductor, a material with special electrical properties that forms the basis of electronic "chips."

"smart machine" a machine or instrument whose activities are controlled by a microprocessor "brain."

sonar a locating system that provides a picture of objects by bouncing sound waves off them.

teaching pendant a small hand-held device resembling a calculator that is used to program a robot.

terraform to convert the environment of another planet or satellite to conditions similar to those of Earth (Terra).

threshold the point at which an effect just begins to occur.

transistor a small block of semiconducting material (such as silicon) that acts as part of an electronic circuit. It may be used as an amplifier (increasing an electrical signal), a detector, or a switch.

ultrasonic pertaining to very high-frequency sound, too high-pitched for human ears to detect.

vacuum tube an electronic device in which a controlled flow of electrons takes place through a sealed glass container from which practically all the air (or other gases) has been removed.

voice synthesizer an electronic device that generates sounds (phonemes) that can be combined to form realistic-sounding words.

For Further Exploration

Books About Robots

Ayers, Robert U. and Steven M. Miller. *Robotics: Applications and Social Implications.* Ballinger: Cambridge, Massachusetts, 1982.

Da Costa, Frank. *How to Build Your Own Working Robot Pet.* TAB Books: Blue Ridge Summit, Pennsylvania, 1979.

Heiserman, David L. *Build Your Own Working Robot.* TAB Books: Blue Ridge Summit, Pennsylvania, 1976.

Heiserman, David L. *How to Build Your Own Self-Programming Robot.* TAB Books: Blue Ridge Summit, Pennsylvania, 1979.

Heiserman, David L. *Robot Intelligence... with Experiments.* TAB Books: Blue Ridge Summit, Pennsylvania, 1981.

Knight, David C. *Robotics.* Wm. Morrow & Co.: New York, 1982.

Krasnoff, Barbara. *Robots: Reel to Real.* Arco Publishing Co.: New York, 1982.

Loofbourrow, Tod. *How to Build a Computer-Controlled Robot.* Hayden Book Co.: Rochelle Park, New Jersey, 1978.

McCorduck, Pamela. *Machines Who Think.* W. H. Freeman & Co.: San Francisco, 1979.

Milton, Joyce. *Here Come the Robots.* Hastings House Publishers: New York, 1981.

Safford, Edward L., Jr. *The Complete Handbook of Robotics.* TAB Books: Blue Ridge Summit, Pennsylvania, 1978.

Susnjara, Ken. *A Manager's Guide to Industrial Robots.* Corinthian Press: Shaker Heights, Ohio, 1982.

Warrick, Patricia S. *The Cybernatic Imagination in Science Fiction.* MIT Press: Cambridge, Massachusetts, 1980.

Winklessless, Nels and Iben Browning. *Robots on Your Doorstep: A Book About Thinking Machines.* Robotics Press: Beaverton, Oregon, 1978.

Periodicals

ROBOTICS AGE: *The Journal of Intelligent Machines*
Strand Building
174 Concord Street
Peterborough, NH 03458
[A bimonthly magazine covering all aspects of robotics, from industrial to hobby, with reports on new advances in robot vision, motion, and speech systems, theoretical articles, discussions of social implications, and "how-to" articles.)

Articles on robots can also be found in science and home computing magazines such as *Creative Computing, Discover, High Technology, Popular Mechanics,* and *Science Digest.*

Organizations

Robot Institute of America
One SME Drive
P.O. Box 930
Dearborn, Michigan 48128
[Emphasis on industrial robots]

United States Robotics Society
1450 Todd Street
Mountain View, California 94040
[For the "garage experimenter"]

Suppliers

Robot Shack
P.O. Box 582
El Toro, California 92630
(Offers kits and parts for robot hobbyists]

Index

Page numbers in italics indicate illustrations.

ACRONYM, 115
Adam and Andrea Android, *104*, 106
Ahmad, 100
Amazing Micro-Mouse Contest, 91, *92*
Androbot, 83, *84, 85*, 96, 106
android, 12, 16
Android Amusement Corporation, 106
arithmetic unit, 23
Armatron, 108, *108*
artificial intelligence (AI), 21, 22, 114, 116, 118, 119
"artificial skin," 29
Asimov, Isaac, 13, 14, 16, 40, 45, 113
Asimov's "Three Laws of Robotics," 13, 14, 16, 113
astronauts, 69, 75
automata, 8, 9, 10, *10*, 11, *11*, 12, 13, 16
automated factory, 55, 56, 112
Avatar, 99, *99*, 101

Balmer, Charles, Jr., 99, 100, 101
Beley, Gene, 106
Bierce, Ambrose, 12
Big Trak, 107, *107*
binary code, 22, 23
"biochips," 120
Blankenship, John, 97, 98, 99
blind, robotic aids for, 35, 64, 65, 120
B.O.B., 83, *84, 85, 85*, 96
bomb squad robot, 49, *50*
Bubble Bot, *6*, 94, *95*
bumper switches, 25, 95
Bumpy, 94, *95*
Bushnell, Nolan, 83, *84, 85*, 106

CAD-CAM system, 56
Capek, Karel, 13
Cardiology Patient Simulator, 57, *58*
chess program, 115
Chess Robot Adversary, 110
Cincinnati Milacron, T^3, *38*, 43, *45*
Clyde the Claw, 37
computers, 18, 21, 22, 23, 25, 26, 29, 34, 36, 47, 56, 60, 61, 67, *67*, 73, 74, 82, 83, 89, 96, 100, 102, 103, 107, 110, 113, 114, 115, 118, 119
 language, 33
 memory, 22, 28
 programs, 23, 59, 60, 113, 118
ComRo I, 88, *89*, 94, *96*
ComRo TOT, 87, *87*, 88, 94, *94*
controller, 47
C-3PO, *15*, 16, 19
cyborg, 120

Dacosta, Frank, 95
Daedalus, 9
Dertouzos, Michael, 118
Devol, George, 40, 42
Dixon, Donald, 100
"Domestic Android," 102, 103, *103*, 104
Draper wrist, 34
Dreyfus, Hubert, 115
Drink Caddy robot, 106
DROID-BUG, 108, *109*
Dr. Who, 16

economics of robot use, 52, 53, 54
Elmer and Elsie, 20
Empire Strikes Back, The, 94
Engelberger, Joseph, 17, 40, 42, 61, 101, 114
entertainment robots, 104
Entropy, 96, *97*
Everett, Bart, 100, *101*

Faber, Joseph, 9
"feedback," 28, 45
"feely bumpers," 82
Fifth Generation Computer Project, *117*, 118, 119
Forbidden Planet, *14*, 16
Frankenstein, 11, 13
Frisina, Tom, 85

General Robotics Corporation, 113
"generations" of computers, 118
Greenblatt, Richard, 115
Grosz, Barbara, 116

Hal, 36
Hamlin, Jerome, 87, 88, 94, *95*
handicapped, robot aids for, 61–67, *63*, *65*, *66*, *67*, 119
Harkins, Arthur, 119
Harvey, 57, *58*, 59
heart patient robot, 57, *58*, 59
Heath, 80, 85
Heath Educational Robot, 80
Hephaestus, 8, 9
Hero I, *6*, *78*, 80, 81, *81*, 85, 86, 88
Hillis, William, 20
Hitachi Limited, 33, 44
"Hi-Ti" hand, 33
Hoffman, E.T.A., 10
home computer, 7, 25, 80, 83, 89, 90, 111
home robot, 7, 8, 79, *81*, 85, 86, 87, 90, 102, 106, 111

126

Hot Tots, 105, *110*
household robot, 16, 20, 85, 87, 88, 89, 94, 103, 111
Huey, Dewey, and Louey, 16
Humanoid Robot, 105

IBM, 43, 44
Iguchi, Toshio, 47, 48
industrial robots, 6, 17, *18*, 19, *32*, 37, 39, 40, 42, 43, *43*, 44, 45, *45*, 46, *46*, 47, 48, *51*, *52*, 53, 54, 55, 56, 81, 86, 90, 113, *117*
input system, 23
Intelledex, 605 robot, 26, 31, *31*, 32
I, Robot, 13, 40

Jacquet Droz, Pierre and Henri-Louis, 9, 11, *11*
Japan, 8, 33, 37, 53, 54, 55, 56, 60, 64, 108, 118, 119
Japanese robots, *52*, *53*, *117*
Jenus (Genus), 88, 89
Jet Propulsion Laboratory, 62, 73, 74
joysticks, 34, 62, 108, *108*

keyboard, 26, 80, 89, 107
keypad, 26, 88
Klatu, 102, 103, *103*
Kurzweil machine, 35, 36, 37

Lally, Eugene F., 109
Learm, 19
Levin, Ira, 119
Loofbourrow, Tod, 93, 95, 101
Lunakhod, 71

MacHack, 115
Maker, the, 45, *46*
manipulators, 17, 18, 26, 33, *41*, 42, 44, 47, 62, 64, 72, 74, 77, 108, *112*
Mars, 70, 72, 73, *73*, 74, *74*, 75, 77
McClees, David, 20
mechanical arm, 110
mechanical devices, 59, 62
medical robots, 59–61
MELDOG, 65
Melkong, *63*, 64
memory storage, 19, 64
memory unit, 22, 23
mental patient, computer simulation of, 115
Messmore, Francis B., 57
microcomputer, 34, 62, 87, 88
microprocessor, 36, 107, 118
Mike ("microtron"), 93, 101
MIT, 20, 116, 118
mobile robots, 34
Mobots, 40
"Monkey and Bananas" problem, 116

moon, 69, 70, 71, 72, 75, 76, 77
movie robots, *14*, *15*, 16
Moxon's Monster, 12, 13
multifunctional, 17, 18
Myers, Dr. John D., 60

NASA, 62, 70, 72, 75, 76, *76*, 107
non-servo-type robots, 45, 46, 47

Octek Inc., 31
Odetics, Inc., ODEX-1, *33*, 34
Oldfield, Gene, 90, 96, 97, *97*
on-board control panel, 108
Optacon reading machine, 67, *67*
Overton, Ken, 55

PARRY, 115, 116
pattern recognizing program, 115
personal computers, 26, 44
personal robots, 16, 64, 80, 81, 83, 90, 96, 109, *109*
"pick and place robot," 46, 47
pixel (picture element), 29, *30*
Poe, Edgar Allan, 10
police robots, 48, 49
Prab model E robot, *38*
program, 23, 29, 46, 59, 60, 80, 110, 116, 118
programming, 26, 88, 89
Psycho, 9
PUMA, 42, 43, *44*
Pygmalion, 8

Quasar Industries, 17, 102, 103, *103*, 104

radio controls, 83
radio signals, 70, 72, 83
Ralph Roger Robot, 104, 105, *105*
RB Robot Corporation, RB5X, 81–85, *82*
rechargeable power supply, 80
Reichelt, Anthony, 17, 102, 103
remote control, 34, 40, *41*, 48, 49, 70, 71, 72, 83, 87, 105, 106, 107
Remote Mobile Investigation Units, 49
reprogrammable, 17, 18
Resusci-Annie, 59
ROBART, ROBART II, 100, *101*
Robbie, *14*, 16
ROBIE (PR-1), 109, *109*
ROBOCYCLE ("Bike Bot"), 96, 97
Robot Basic, 26
robot
 cart, 64
 chess player, 114–115
 definition of, 13, 17–18, 37
 "diagnostic cabinet," 61
 doctor, 59, 60
 "dog," 94, 95, *96*
 hospital aids, 64

127

kits, 108
landers, 72
lunar rover, 71
mail carriers, 48
manipulator, 62
manufacturers, 43–45
"marriage," 119
memory, 64
mice, 21, 91, *92*, 93
monsters, 16
myths, 8, 9
patients, 59
"rights," 120
"seeing eye dog," *65*
sheep shearer, 49
spacecraft, 69, 70
speech, 36
teaching tool, 59
toys, 107, 108, 110
travel machine, 84, 85
waiters, 48, 112
watchmen, 48, 100
workers, 76, 113
worms, 20
Robot Factory, The, 104
Robot Institute of America, 17, 18
Robot Repair, 90, 97
Robot Shack, 108, 109, 110
Robotic Aid, 62, 64
robotic intelligence, 74
Robotics Age, 97, 99, 107, 108
Robotics International Corporation, 88
robots
  advantages of, 50
  attitudes toward, 52, *112*, 113, 114
  limitations of, 55
  maintenance of, 39
  reprogrammable, 40
  resistance to, 51, 52, 55
  sabotage of, 51
R2-D2, *15*, 16, 19, 48, 80, 114
R.U.R., 13, 19

self-controlled robots, 94
self-reproducing robots, 76, 77
senses, artificial, 47
senses, robot, 28
sensors, 28, 36, 45, 48, 55, 62, 74, 82, 83, 89, 97, 100, 109, 118
sentry robots, 48
servo-controlled robots, 45, 46, 47
servo-mechanism, 45
Shakey, *24*, 25, 26
Shannon, Claude, 21
Shelley, Mary Wollstonecraft, 11, 12
show robots, 17, 88, 94, 102, *103*, 104, *104*, *105*, 107, *110*
Silent Running, 16
silicon chips, 22, 25, 107, 118
Sim One, 59

Six Robot, 104
"smart machines," 67, 111, 118
Snoopy, 48, 49, *49*
software, 100
sonar ranging system, 83, 96, 102
Sonicguide, 64, *66*
Soniscan system, 95
space, robots in, 69–71, 73–74, 75, 76
Space Shuttle, 42, 69, 75
Spectrum magazine, 91
speech synthesizers, 35, 36
Stanford Research Institute (SRI), 24, 25, 44, 116
Stanford University, 60, 62, 115
Star Trek, 16
Star Wars, *15*, 16, 80
Stepford Wives, The, 119
Susnjara, Ken, 39, 40, 45
synthesized robot voice, 36, 99, 109

talking machines, 35, 36
teaching pendant, 26, 27, 43, 80
teaching robot, 81
"terraform," 77
Terrapin, Inc., 20
TIMEL, 97, 98, *98*, 99
Time Magazine, 7
tooling, 47
Topo, 83, *84*
toy manipulator, 108, *108*
transistor, 25, 118
Turk, 9, 10, *12*, 110
Turtle robots, 20, *21*
TV camera, 25, 26, 40, 65, 73

ultrasonic ranging system, 80, 87, 89, 93
undersea robots, 50, 77
Unimate, 42, 43, 86, *86*
Unimate 4000B, 37
Unimation, 17, 42, 43
Unimation PUMA-250, 62
United States Robots, 44, 45, *46*

vacuum tube, 23, 25, 118
Viking landers, 72, 73
vision devices, 65, 67
vision systems, robot, 29, *30*, 31, *31*, 32, *32*, 44, 64, 65, 73, 74, 111
vision system, "real-time," 32
voice recognition system, 28, 93, 112
voice synthesizer, 35, 36, 62, 67, 80, 83, 84, 95, 110
von Kempelen, Baron Wolfgang, 9, 10

Walter, William Grey, 20
Wilson, Patrick, 116
Wires, 94

X-1, 108, 109, *109*

128